Annie Shalom Samji Isaac Chandra

Intelligent Freeform Deformation for LED Illumination Optics

SPEKTRUM DER LICHTTECHNIK **BAND 16**

Lichttechnisches Institut
Karlsruher Institut für Technologie (KIT)

Intelligent Freeform Deformation for LED Illumination Optics

by
Annie Shalom Samji Isaac Chandra

KIT Scientific Publishing

Karlsruher Institut für Technologie
Lichttechnisches Institut

Intelligent Freeform Deformation for LED Illumination Optics

Zur Erlangung des akademischen Grades eines Doktor-Ingenieurs von der KIT-Fakultät für Elektrotechnik und Informationstechnik des Karlsruher Instituts für Technologie (KIT) genehmigte Dissertation

von M.Eng. Annie Shalom Samji Isaac Chandra geboren in Kolachel, Indien

Tag der mündlichen Prüfung: 16. November 2017
Referent: Prof. Dr. Cornelius Neumann
Korreferent: Prof. Dr. Wilhelm Stork

Impressum

KIT Scientific Publishing

Karlsruher Institut für Technologie (KIT)
KIT Scientific Publishing
Straße am Forum 2
D-76131 Karlsruhe

KIT Scientific Publishing is a registered trademark of Karlsruhe Institute of Technology.
Reprint using the book cover is not allowed.

www.ksp.kit.edu

Print on Demand 2018 – Gedruckt auf FSC-zertifiziertem Papier

ISSN 2195-1152
ISBN 978-3-7315-0741-3
DOI 10.5445/KSP/1000077210

Intelligent Freeform Deformation for LED Illumination Optics

Zur Erlangung des akademischen Grades
DOKTOR-INGENIEURS

von der Fakultät für

Elektrotechnik und Informationstechnik
des Karlsruher Instituts für Technologie (KIT)

genehmigte

DISSERTATION

von

M.Eng. Annie Shalom Samji Isaac Chandra

geb. in: Kolachel, Indien

Tag der mündlichen Prüfung: 16.11.2017

Hauptreferent: Prof. Dr. Cornelius Neumann
Korreferent: Prof. Dr. Wilhelm Stork

Kurzfassung

Durch die neuen Möglichkeiten der Freiformoptik fordert die Beleuchtungsindustrie anspruchsvolle Designtechniken und neuartige Optimierungswerkzeuge für die Optikentwicklung. Durch die Bereitstellung eines fortschrittlichen und benutzerfreundlichen Software-Frameworks kann die Arbeit des Optikentwicklers vereinfacht und der bisher notwendige Erfahrungsschatz bei der Optikentwicklung reduziert werden, was das Hauptziel dieser Arbeit ist. Bei der Freiformoptik ist die Optimierung aufgrund einer Vielzahl von Parametern limitiert. Diese Einschränkung wurde durch die Verwendung einer indirekten Technik, die als Optimierung mit Freiform Deformation (OFFD) bekannt ist, aufgehoben. Obwohl diese Technik für viele Beleuchtungsaufgaben gut funktioniert, bringt sie viele Herausforderungen mit sich. Dazu gehören die Auswahl geeigneter Kontrollpunkte des kubischen Gitters, die Analyse zu Herstellbarkeit, die Auswahl des Optimierungsalgorithmus und die lokale Modifikation der Oberflächen. Diese Herausforderungen werden in der vorliegenden Arbeit erarbeitet und für verschiedene Beleuchtungssituation validiert. Die Techniken wurden so umgesetzt, dass sie eine Intelligente OFFD bilden, die die Arbeit des Optischen Designers erleichtert. Dieser muss lediglich noch den Initialentwurf und die photometrische Anforderungen definieren und erhält als Ergebnis eine herstellbare und effiziente Freiformoptik. So wird eine automatisierte Optikentwicklung ermöglicht die zur Entwicklung anspruchsvoller Freiformoptiken in der Beleuchtungsindustrie beiträgt.

ABSTRACT

Due to the revolution of the freeform optics in the field of illumination, the lighting industry requires sophisticated design techniques and novel optimization tools. By providing an advanced yet user-friendly software framework, the need to rely upon an optical designer's expertise can be reduced which is the main aim of this work.

In freeform optics, the optimization is limited due to a large number of parameters present in it. This limitation was overcome by using an indirect technique known as optimization using freeform deformation (OFFD). In this approach, the optical surface is placed inside a cubical grid. The control vertices of this grid are modified which deforms the underlying optical surface during the optimization. Though this technique proved to work well for many illumination tasks, it has left behind many challenges to the optical designer. These challenges include selection of appropriate control vertices of the cubical grid, manufacturing feasibility analysis, selection of suitable optimization algorithm and local modification of surfaces to improve the optical performance. These challenges are addressed in this dissertation by providing suitable mathematical design techniques and they are validated for different illumination design tasks to show its versatility and its potentiality. This implementation transformed the OFFD into an intelligent tool replacing the optical designer's efforts in terms of his decision making during the entire design process.

CONTENTS

CHAPTER 1

INTRODUCTION AND TERMINOLOGY

This chapter introduces the terms necessary to understand the freeform optics design in modern illumination systems. It addresses the motivation behind this research and outlines the organization of the dissertation.

1.1 INTRODUCTION

Light impacts the life on the earth. Light is a form of electromagnetic energy that comes from the Sun as a natural form. As man wants to get rid off the darkness to double his activities, he began to create artificial sources of light. This paved the way for the invention of artificial light sources one after the other. Artificial light sources that gained popularity include incandescent bulb, gaseous discharge lamps and Light Emitting Diode (LED). Compared to all these sources, LEDs have revolutionized and fascinated the whole world during the last decade by its attributes like energy efficiency, heat-light conversion ratio and longer life time. Due to these advantages, it has found its attention in all sectors ranging from daily use home luminaries, decorative lights, street lights, automotive headlights to lettuce cultivation in the International Space Station. Though LEDs are used in diverse applications, there are certain tasks which require high efficiency, uniformity and glare elimination. In order to address these requirements, suitable optics must be designed in front of them to collect and distribute the light where it is needed. This field of collecting and distributing the light effectively is called as illumination. Illumination deals with an effective transfer of light from the source to the target [1]. The effective transfer of light can be attained using optical elements which come in different types.

There are many varieties of optics that can be used in the illumination design. They are refractive optics (lenses), reflective optics (mirrors), total internal reflection optics (light guides), scattering optics (diffusers) and hybrid optics [1]. Based on the application's needs, individual or combination of optics can be used. For example, compact optics can be built using refracting optics where a wide spread of light can be attained using reflective optics. Scattering optics provide uniformity at the expense of efficiency.

Once the type of the optics is selected, a suitable optical design must be carried out satisfying the lighting requirements. This includes making sketches of the optical system, generating mathematical descriptions of the element and convert into computer-aided design (CAD) format. The optical design community has grown increasingly with the advancements in the CAD and graphics industry. It has progressed from simple curves, conics, aspherics and slowly developed into freeform surfaces. Any surface can be considered as freeform as long as its mathematical description does not rely on the assumption of any symmetry [2]. Their freeform nature provides additional degrees of freedom to create efficient optics with fewer elements [3].

This popularity is further enhanced by the growth of the manufacturing technologies that ranges from prototyping to tooling for the mass production of the freeform surfaces. Due to these reasons, freeform optics evolve at a rapid rate making a greater impact in the field of illumination. This evolution has brought many challenges on its way as well mainly to develop efficient design and optimization techniques for many illumination applications. This aspect on the freeform optics to develop efficient design and optimization techniques will be addressed in this dissertation.

1.2 MOTIVATION

The revolution of the LEDs have also led to the increase in its size in the order of some centimeters in the recent years. Some examples include chip on board (COB) LED and LED light engines. Therefore considering the dimension of the source at an early stage is critical while designing compact optics. There are many design methods available to date but most of them assume the light source as a point and have to be solved using complex differential equations. This is because finding a suitable ray transformation to yield a continuous refractive or reflective surface for complicated LED models is difficult.

Therefore optical designers always have a great interest in designing efficient, robust optimization algorithms to improve their initial design. Developments in simulation programs, optimization algorithms and computational capabilities have brought their interests close to the reality but they did not solve all the issues leaving optimization still a challenging task.

Some of the difficulties faced during the optimization of freeform surfaces are selecting a suitable optimization algorithm, modification of surfaces based on the photometric requirements, dealing with a large number of optimization variables that have no relationship with lighting requirements and is time intensive.

The limitation on dealing with large number of parameters present in the NURBS was overcome by using a technique known as optimization using freeform deformation (OFFD) proposed by Wendel [4]. This method uses indirect optimization by placing the optical surface inside a cubical grid. Instead of directly modifying the optical surface containing thousands of parameters, the vertices of the cubical grid are modified which deforms the underlying optical surface. Due to this global transformation, this approach yields continuous and smooth surfaces. But unfortunately, the surfaces created by them are not always feasible to manufacture because of its deformation.

Moreover, another challenge in this technique is the selection of appropriate control vertices of the grid. This is because the control vertices share no relationship with the optical performance. When irrelevant control points are selected, the computational complexity increases. These difficulties make OFFD still a challenging task and require the need of an expert optical designer to rightly modify the surface in the desired direction. These challenges can be solved only if the intelligence of the optical designer is transferred to the routine.

This dissertation aims to solve the challenges present in the Optimization using freeform deformation technique thereby making it more advanced. These challenges can be solved only by designing suitable mathematical techniques. So this work aims to implement new design techniques in the OFFD and analyze its performance. The implemented techniques must be robust and should not confine to a particular illumination design problem.

These techniques must improve the speed, optical performance and should demand minimal efforts from the optical designer. At the end, the OFFD should be more intelligent, autonomous and extensible thereby, pushing the illumination optics one step further ahead.

1.3 DISSERTATION OUTLINE

Chapter 2 provides an overall view of the existing optical design and optimization techniques in the field of illumination and its current challenges. Chapter 3 details the optimization using freeform deformation technique. This chapter ends by introducing the design examples which will be used throughout the dissertation.

The previous work [4] employed Nelder-Mead simplex optimization algorithm in the OFFD. But many researchers have studied on it and proposed different variants of it. So these different variants are investigated, implemented and its performance was analyzed in the OFFD. Then, genetic algorithm which is based on global search technique was studied and implemented. The exploration of these algorithms and comparison study in terms of speed and optical performance are given in the chapter 4.

The freeform surfaces are commonly represented as non-uniform rational basis splines commonly known as NURBS. Global deformation can be attained with NURBS based OFFD. But there are certain designs which require local deformation of the optical surfaces. For such design cases, the NURBS cannot offer best results due to its underlying surface representation. An alternate surface representation called T-Splines can make this possible. But this has neither been applied to any optimization system so far nor its optical performance has been analyzed. So T-Splines have been implemented and its impact on the OFFD is analyzed. More technical description and implementation details are found in the chapter 5.

While designing compact optics, the most limiting factor is the consideration of manufacturing tolerances during the design process. So chapter 6 details the implementation of manufacturing feasibility analysis in the OFFD and explains how it obtains manufacturing feasible surfaces in short duration.

Chapter 7 shows how the entire OFFD can function on its own without any manual intervention. To attain this autonomy, the intricacies of the OFFD and the relationship of the OFFD grid with the target light distribution has to be studied. Previously, this relationship was not known and the OFFD relied heavily on the knowledge of an expert designer. This chapter provides design techniques to establish this much-needed relationship. By doing so, the work of an optical designer or his intelligence is transferred to the OFFD. A comparison study is made between the previous work and the newly implemented design techniques and the results are discussed at the end. Finally, the dissertation ends with a research summary and provides an outlook for further improvements in chapter 8.

1.4 TERMINOLOGIES

SURFACE REPRESENTATION

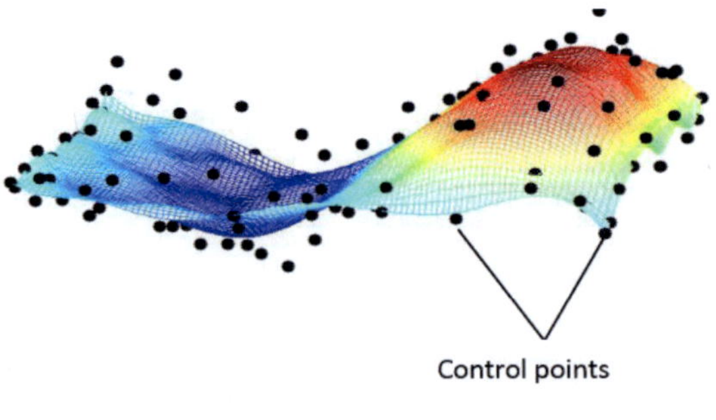

Figure 1.1: NURBS patch with the control points distributed along the surface.

The freeform surfaces are mathematically represented using non-uniform Rational B-Splines commonly known as NURBS. NURBS techniques are used in computer-aided systems as well as in raytracers. The figure 1.1 shows an example of a NURBS patch with control points being distributed along the surface. Freeform surfaces require 20 to 1000 control points to represent its surface accurately. A NURBS geometry is defined by its degree, control points, knots and basis functions. NURBS surface is the parametric tensor product surface and is defined using the equation 1.1.

$$\boldsymbol{S}(u,v) = \frac{\sum\limits_{i=0}^{n}\sum\limits_{j=0}^{m} w_{i,j}\boldsymbol{P}_{ij}N_i^p(u)N_j^q(v)}{\sum\limits_{i=0}^{n}\sum\limits_{j=0}^{m} w_{i,j}N_i^p(u)N_j^q(v)} \tag{1.1}$$

where $\boldsymbol{P}_{ij}$ is $(n+1)\times(m+1)$ rectangular array of control points, $w_{i,j}$ are its weights. $N_i^p(u)$ and $N_j^q(v)$ are the basis functions of degree p and q in u and v directions respectively associated with the knot vectors represented by

$$\overrightarrow{\boldsymbol{u}} = [u_o, u_1, \ldots, u_r]\,,\, \overrightarrow{\boldsymbol{v}} = [v_o, v_1, \ldots, v_s] \tag{1.2}$$

where $r = p + n + 1$ and $s = m + q + 1$.

PHOTOMETRIC QUANTITIES

In order to describe and evaluate illumination systems, fundamental photometric quantities are necessary [5]. They are given by

Luminous flux ϕ, total radiated power as perceived by human eye whose unit is *lumen* denoted by the symbol *lm*.

Luminous intensity $I = \frac{d\phi}{d\Omega}$, luminous flux (ϕ) per unit solid angle (Ω) whose unit is *candela* denoted using the symbol *cd*.

Illuminance $E = \frac{d\phi}{dA}$, luminous flux per unit area (A) whose unit is $lumen/m^2$or lux.

Luminance $L = \frac{d^2\phi}{d\Omega dAcos\theta}$, luminous flux per unit solid angle per unit projected area perpendicular to the specified direction whose unit can be represented as $candela/m^2$.

RAYTRACING

To analyze the freeform surfaces represented as NURBS, raytracing has to be performed. Illumination systems often have multiple or complex light sources which are modelled as rayfiles containing millions of rays. The detector is modeled as an angular or surface sensor based on the requirements. All the physical properties like Fresnel effects, absorption, optical materials and coatings can be modeled and are included in the simulation model.

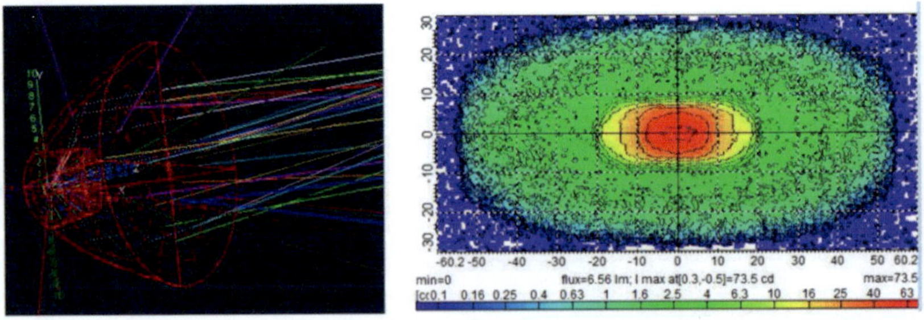

(a) (b)

Figure 1.2: An optical simulation setup using *Lucidshape* [6] showing (a) the rays created from the LED traveling through various paths of the optics and (b) the detector where the intensity is being calculated.

The raytracer shoots the rays and these rays travel until it meets any interface as there is no pre-defined path for its traversal. When it hits any surface, the deviation of the ray path is calculated using the laws of refraction or reflection and it reaches the detector. In the end, the sum of all contributions either as intensity or illuminance is stored in the pixels based on the resolution of the detector. This type of raytracing to understand the behavior of the optics is known as non-sequential raytracing. Some examples of the commercial non-sequential raytracers are *FRED* from *Photon Engineering* [7], *Lucidshape* by *Synopsys* [6] and many others. The figure 1.2 shows an optical simulation setup and shows how the rays travel along the optics until it reaches the detector where the intensity is being calculated. The results from the raytracing help to analyze wide range of illumination characteristics including luminous intensity distribution, zonal luminous flux calculation, illuminance and many others. These illumination characteristics can be described effectively using the photometric quantities which will be addressed next.

Important Factors in Illumination Design

The two important parameters used to evaluate an illumination system are optical efficiency and uniformity [1].

Optical efficiency (η) is defined as the ratio of the total flux at the target ϕ_{target} to the flux emitted from the source ϕ_{source}.

$$\eta = \frac{\phi_{target}}{\phi_{source}} \times 100\% \tag{1.3}$$

Uniformity defines how the simulated or measured distribution agrees with the target distribution. The target distribution can be either intensity, illuminance or luminance. Common methods used to analyze uniformity are the root mean square (RMS) deviation, peak to valley variation of the distribution and many others.

This chapter provided the general introduction, motivation behind this work, outline of the dissertation and basic terminologies required in the field of illumination optics design. The coming chapters provide in-depth detail of the related work, limitations of them and how this research has improved some of them thus expanding the boundaries of the illumination design.

CHAPTER 2

RELATED WORK

The growth of freeform optics is dramatic during the last ten years. This growth is always confronted with many challenges along the way right from the inception to its maturity till today. This chapter takes us through this journey by grazing some of the common techniques used to create freeform optics and its challenges which motivate this dissertation.

2.1 DESIGN TECHNIQUES IN ILLUMINATION

The direct design techniques generate optical surface based on the analytical techniques without any iteration when the target prescriptions are given [1]. Some of the most commonly used techniques in the field of illumination are explained in this section briefly.

2.1.1 TAILORING

Tailoring of optical freeform surfaces for point sources were introduced and made popular by Muschaweck [8] and Oliker [9]. In this method, the design task is formulated into a set of partial differential equations (PDE) which are then solved using standard numerical methods. The solution of the PDEs yields the shape of a single optical surface, which can be a freeform mirror or a freeform lens surface. This technique allows us to exactly calculate the optical surface that could transform the intensity distribution of a given source into a specified target distribution as shown in the figure 2.1. The angular source characteristics represented as $(\theta_0, \theta_1, \ldots, \theta_4)$ are first mapped to the target intensity points $(y_0, y_1, \ldots, y_4)$. Then these properties are translated into PDEs and the corresponding normal vectors $(\mathbf{N_0}, \mathbf{N_1}, \ldots, \mathbf{N_4})$ are calculated. Once this is known, a contour of the surface containing the points $(P_0, P_1, \ldots, P_4)$ with a minimum source-optics distance r_{min} can be constructed. This tailoring is used by wide varied of optical designers and they could attain illuminance or intensity requirements which were impossible before.

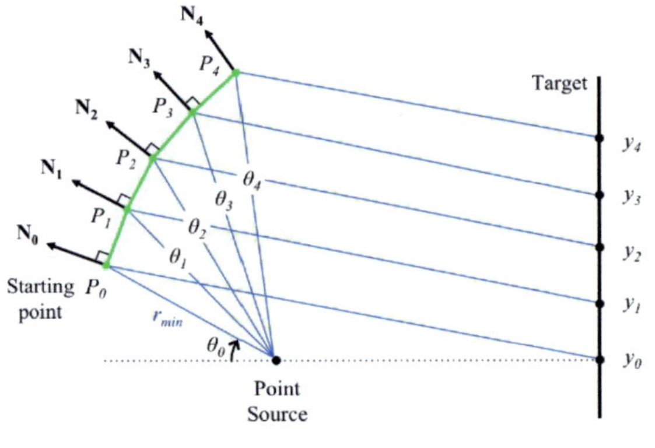

Figure 2.1: Freeform reflector curve (green) constructed using tailoring method [10].

2.1.2 Simultaneous Multiple Surfaces (SMS) Method

In this method, the desired lighting requirements can be attained with multiple surfaces as they provide more degrees of freedom. In the tailoring method described in section 2.1.1, the location of the target for each incoming source ray is first defined. Based on this, the surface normal is calculated according to the law of refraction or reflection. But there is no definition when a second ray hits at the same surface point [10]. As a consequence, we do not have full control over the obtained lighting distribution.

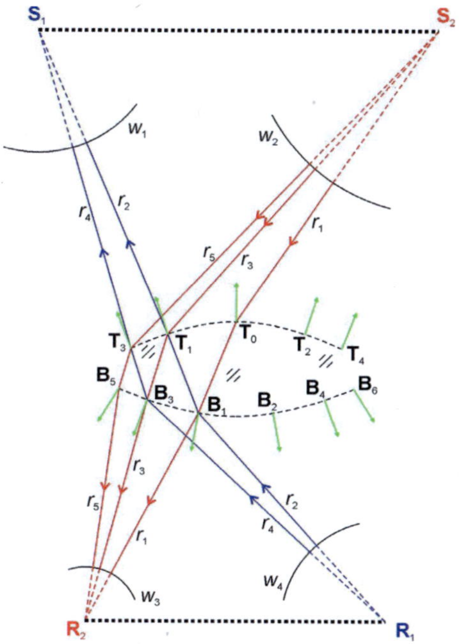

Figure 2.2: Construction of surfaces using SMS method [11]

This problem is solved by constructing multiple surfaces simultaneously for extended sources using a recent popular method called SMS method [12]. This method allows to control up to three orthonormal bundles of rays [1]. By using two surfaces, the two input wave fronts are coupled with the two output wave fronts [10]. This gives more control on the input-output optical characteristics. The figure 2.2 shows an example of how the points $(T_0, T_1, \ldots, T_4)$ and $(B_1, B_2, \ldots, B_6)$ for two surfaces are calculated. The edge rays $(r_1, r_2, \ldots, r_5)$ of the source $\mathbf{S_1S_2}$ are mapped to the target region $\boldsymbol{R_2R_1}$. It is seen that the rays r_2 and r_3 from different source areas meet at the same surface location T_1. But this method uses two surfaces efficiently to redirect these two rays r_2 and r_3 to R_1 and R_2 respectively. By doing so, the optics gains more control on the light distribution. As the surface points are calculated on a point by point basis, solving any complicated expression is not required in this method [3]. Some of the optical design tasks solved using this method include LED headlamps [11] and condenser using two mirrors [13]. But this method requires a complex choice of input and output wave fronts to generate continuous surfaces [10].

2.1.3 Equi-flux Grid Methods

This method was developed by many [10, 14, 15] researchers for different target requirements. In this method, the shape of the optical surface is obtained by mapping the source rays to the target. This is done by comparing the cumulative flux distributions of the source and the target. The intensity distribution of the source $I(\theta, \phi)$ is divided into solid angles (θ, ϕ) with each comprising the same amount of flux as seen in the figure 2.3. Similarly, the target illuminance distribution $E(x, y)$ must be divided across x and y too. Then a mapping will be established between the equi-flux source grid and the equi-flux target as described in the equation 2.1. But this method does not consider the spatial extension of the source and rely on compensation or optimization techniques at the end.

$$\int_{\phi_0}^{\phi} \int_{\theta 0}^{\theta} I(\theta, \phi) sin\theta d\theta d\phi = \int_{x_0}^{x} \int_{y0}^{y} E(x, y) dx dy \tag{2.1}$$

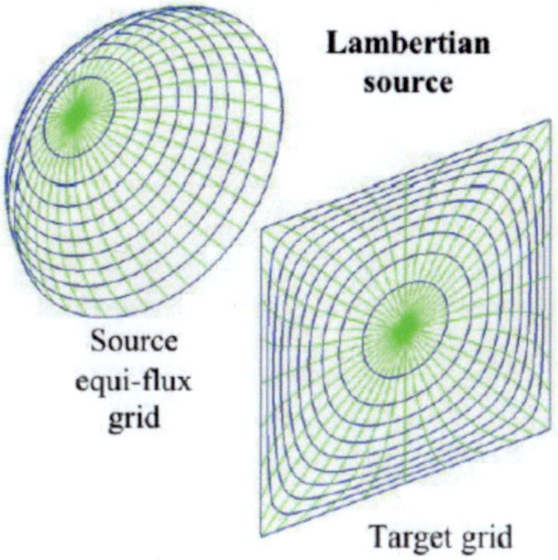

Figure 2.3: Source-target mapping of a Lambertian source to a square target by splitting the source and the target luminous flux into equally spaced grids [10].

2.2 ITERATIVE TECHNIQUES IN ILLUMINATION

The optical design task based on point source assumption does not hold good when the optics size is so compact almost comparable to the size of the LED. To deal with such cases, flux compensation technique was developed to compensate the size of the LED by performing several iterations. In this approach, the ratio between the desired and the actual distribution is first calculated [16]. Then this compensated distribution calculated with the extended source is used when updating the design. The adjustment is usually repeated multiple times until target requirements are completely met [17]. This entire procedure is explained using a flow chart in the figure 2.4. This approach is effective when the maximum size of the source images on the target is less than one fifth of the target extent [18]. When the maximum size of the source images becomes too large, the one-to-one correspondence between the source and the target is broken and the compensation approach becomes less effective [10].

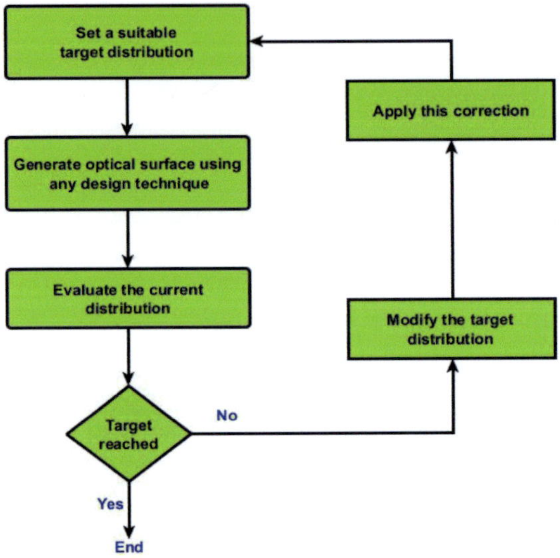

Figure 2.4: Flowchart showing the iterative technique [10].

DESIGN EXAMPLE

A design example incorporating some of these techniques elaborated in the previous sections is presented here to show their significance and limitations of design techniques. In this example, a compact LED flashlight fulfilling two different intensity requirements that could be used in explosive atmospheres was designed and validated [17]. The initial design was made using the tailoring technique described in the section 2.1.1 followed by iterative flux compensation approach explained in the section 2.2. The two different target requirements are realized using TIR hybrid optics by moving the optics along the optical axis as shown in the figure 2.5. The results are validated and the measured results showed an overall agreement with the simulated ones. As shown in the figure 2.6, when the LED is moved forward, ring distribution is obtained and with a backward shift of 1.75mm along the optical axis, collimated distribution as specified in the DIN 14649 standard [19] is obtained. The overall diameter of the optics is 29 mm and 10 mm in thickness and the figure 2.7 shows both distributions being projected on the screen.

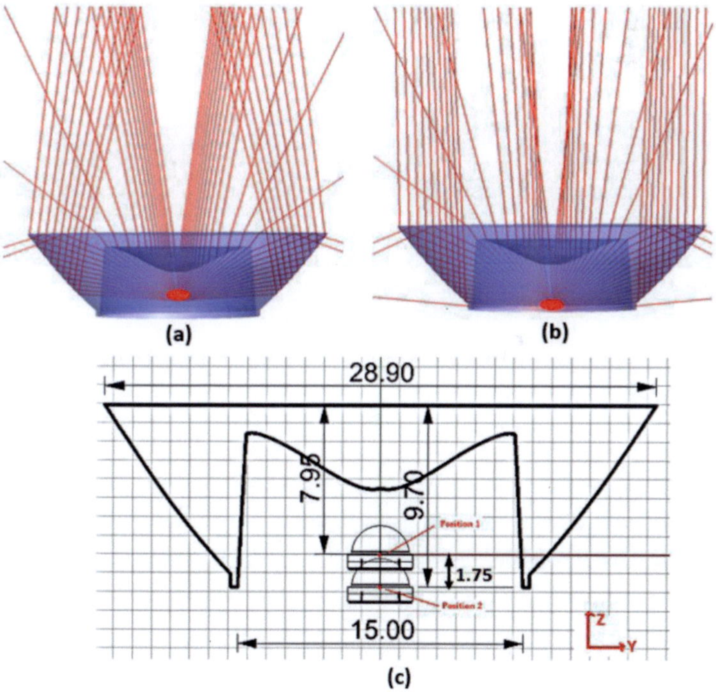

Figure 2.5: TIR hybrid optics with a schematic view of rayfan showing (a) LED at forward and (b) LED at backward position and (c) cross-sectional view with dimensions in millimeters.

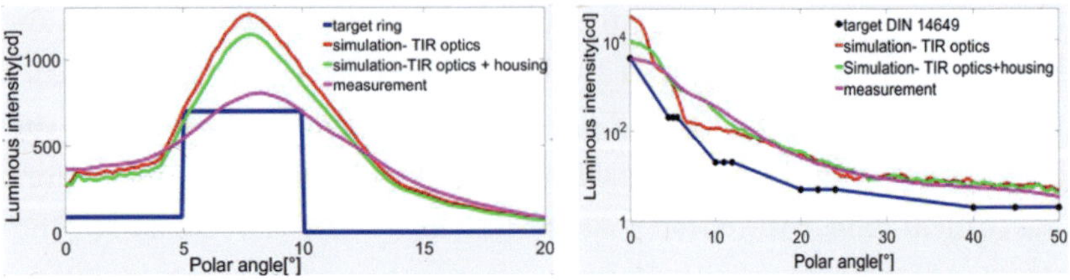

(a) (b)

Figure 2.6: Light intensity distributions showing the expected target distribution, simulation results with only TIR optics, TIR optics along with housing and measured results with (a) showing the ring distribution to be used in fog or smoke environment and (b) showing the distribution verifying the DIN 14649 standard [19].

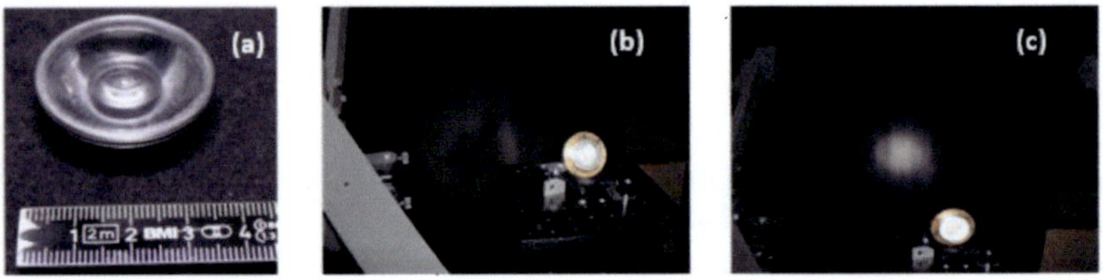

Figure 2.7: (a) prototype of TIR hybrid optics and its measured results projected on the screen (b) for the ring distribution and (c) for the DIN 14649 distribution [19].

So this example showed that the design techniques can be used to create compact optics for any two different light distribution. Since automated design techniques to realize both distributions at a same time are not available, heavy iterative manual tweaking processes were involved as a part of this design. Further development in automation in design processes are needed to achieve faster and better results. Additionally, imparting tool tolerances right from the start of the design helps to avoid deviation in the final results. The limitations in these design techniques can be mitigated only by using an efficient automated optimization system. This system must be capable of understanding the needs of optical designer, converts them into suitable design requirements and creates optical surface satisfying the photometric requirements and manufacturing feasible surfaces in a short span of time. The role of optimization in illumination and its challenges are addressed in the coming section.

2.3 OPTIMIZATION TECHNIQUES IN ILLUMINATION

A common method to enhance the performance of any initial system is to apply standard optimization techniques over it. In the same way, the optical performance of the illumination system can be improved by applying optimization algorithms on the initial surface which is usually generated by one of the design techniques mentioned in the section 2.1. But the potentiality of the optimization can be harnessed only when the illumination problem is appropriately described. This way of describing the illumination requirements is called as merit function and this function governs the optimization variables acting upon it. The optimization variables are the ones which are responsible for the modification of the optical surface and the photometric measure of this modification will be evaluated using the merit function. This step is repeated several times until the desired photometric requirements are met. The main role of the optimization algorithm is to search this optimal solution faster by either maximizing or minimizing the merit function based on the demands of the application.

There are numerous optimization algorithms ranging from very simple to the ones that could mimic nature so closely like genetic algorithms and ultimately leads to machine learning techniques which are already revolutionizing the globe with its potentialities. Each optimization technique has its own merits and its drawbacks and there is no single algorithm so far that could cater all the needs of an optical design.

Optimization was predominantly used in imaging optics but now has become a part of the illumination design since Cassarly introduced it [20, 21]. Koshel and others used Nelder Mead Simplex method to optimize their initial designs and have showed promising results [22, 23, 24]. Slowly other optimization algorithms like genetic algorithm and various evolutionary techniques are being utilized in the illumination design tasks as well [25, 26]. Despite its proven advantages, the growth is still slow. The reason for the slow adoption of optimization techniques in illumination is due to the following factors.

MORE OPTIMIZATION VARIABLES

Traditionally, any optimization algorithm acts directly on the optical surface. This can be achieved either by varying the diameter or thickness of the optics. This is straight forward for simple surface representations like spherical or aspherics but not for freeform optics. As freeform optics is most commonly represented using NURBS, the possible optimization variables could be the control points or its weights, its degree or even the knot vectors. This imposes large parameter space for optimization but hard to manipulate as they share no relationship with the optical performance. This is because the NURBS are initially developed to describe the automotive or aircraft body parts and because of its advantages, it has been slowly adopted to describe the optical surfaces.

Most commonly, the freeform surfaces are modified by changing the control points. So an optimization of freeform optics basically means transformation or displacement of the control points present in the

NURBS to create a new surface, evaluate it using a defined merit function and iterate until the target functionality is attained.

But attaining them is not as simple as one could think of because NURBS surface has hundreds or thousands of control points in three dimensions. Moreover, the optical designer is unaware to operate on the appropriate control points to meet his target requirements. Still, if he optimizes the NURBS directly, it leads to kinks, discontinuities and perturbations which cannot be manufactured at the end [27].

TIME INTENSIVE

The number of rays to analyze the performance of illumination optical system is more compared to imaging optics. The illumination optics which is non-sequential require millions of rays to accurately model all the effects like absorption, Fresnel effects and so on. As the number of rays increases, the raytrace duration also increases proportionally. And also, if the merit functions in the optimization system are not accurately described, the convergence rate is slow leading to longer duration. Though optimization in illumination is an arduous task, this cannot be neglected as there is no direct method to create optical surfaces for extended source available. So in order to effectively use optimization in freeform surfaces, reparameterization of surfaces is essential to reduce the number of optimization variables. This is achieved using a method known as optimization using freeform deformation [4] which is elaborated in the chapter 3.

CHAPTER 3

OPTIMIZATION USING FREEFORM DEFORMATION (OFFD)

Freeform surfaces that meet target lighting requirements can be generated using an advanced method called optimization using freeform deformation. This method couples the freeform deformation technique with optimization routine. The significance of this method is that it uses fewer optimization variables and attain desired optical performance by undergoing global deformation. This chapter explains this technique in detail and proposes how this technique could be improved further.

3.1 STATE-OF-THE-ART

The OFFD tool is capable of attaining the desired optical performance for symmetric as well as for asymmetric systems. This method works well even when the performance of the initial system is far from the target light distribution. It can be directly implemented to the extended LED sources as well.

3.1.1 OFFD PRINCIPLE

OFFD method employs freeform deformation (FFD) technique proposed by Sederberg [28] coupled with an optimization routine. The relationship between the grid and the optical surface is established using the FFD algorithms [28]. The figure 3.1 shows the cubic grid with an optical surface placed inside before and after deformation.

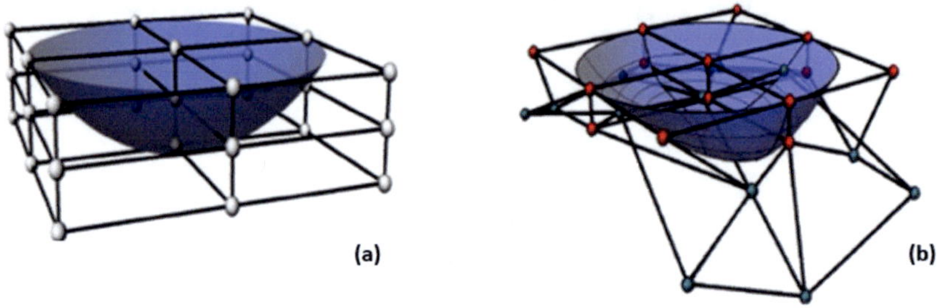

Figure 3.1: 3x3 OFFD cubical grid enclosing the optical surface (a) before deformation and (b) after deformation [4].

The workflow is represented using a flowchart in the figure 3.2. It begins with an input surface whose optical performance needs to be improved. The optical surface is placed inside the grid whose vertices are known as grid control points. The designer could select any combination of grid control points $P_1, P_2, \ldots, P_n$ and this combination is provided as variables to the optimization algorithm. The optimization algorithm has a wide search space for the selected combination and is free to provide displacements along all the three directions of the enclosed grid. As the enclosed grid undergoes a change, the underlying optical surface changes. The deformed surface is then photometrically evaluated by raytracing and based on this result, the optimization algorithm decides the amount of shift for the OFFD grid points in the next iteration. This continues again and again until the target lighting requirements are met. The fundamental building blocks involved are deformation technique, optimization algorithm, raytracing and the merit function governing them. These four blocks essential to understand OFFD are explained in detail in the coming sections.

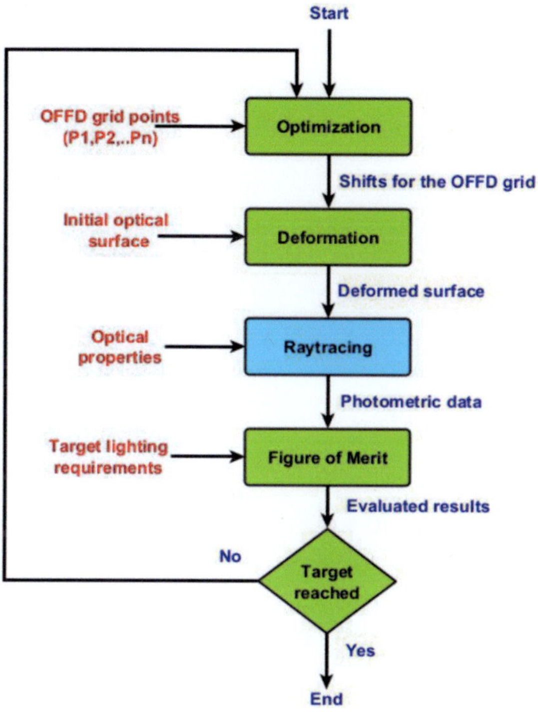

Figure 3.2: Workflow of the optimization using freeform deformation technique (OFFD).

3.1.2 FREEFORM DEFORMATION TECHNIQUE

The main role of modifying the shape of the optical surface is achieved using the freeform deformation (FFD) technique [28]. In this method, the optical surface is placed in a parallelepiped grid. The vertices of the grid are then pushed in or pulled out and this effect is transmitted to the underlying optical surface [29]. This global single line statement is translated into following mathematical steps.

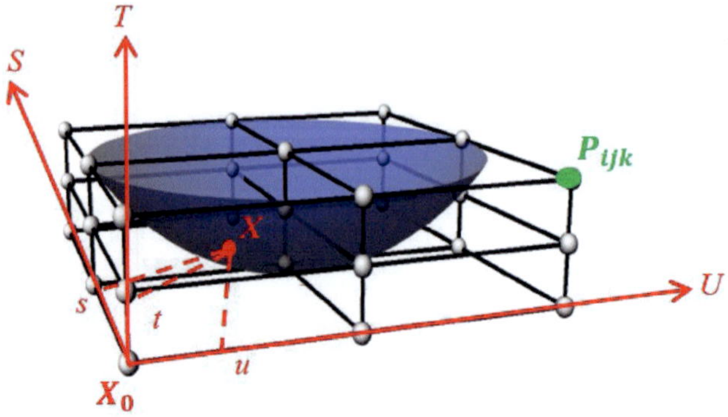

Figure 3.3: Representation of local coordinates (s, t, u) of point $\boldsymbol{X}$ of the optical surface in the 3D OFFD grid space.

As a first step, the optical surface has to be mapped to a rectangular lattice space by a coordinate transformation. This is defined in terms of a tensor product trivariate Bernstein polynomial.

Any point $\boldsymbol{X}$ on the embedded optical surface has local coordinates (s, t, u) as shown in figure 3.3. The expression of local coordinates (s, t, u) of a point $\boldsymbol{X}$ is expressed in equation 3.1

$$\boldsymbol{X} = \boldsymbol{X}_0 + s\boldsymbol{S} + t\boldsymbol{T} + u\boldsymbol{U} \tag{3.1}$$

The vector solution of s, t and u can be found as,

$$s = \frac{\boldsymbol{T} \times \boldsymbol{U}.(\boldsymbol{X} - \boldsymbol{X}_0)}{\boldsymbol{T} \times \boldsymbol{U}.\boldsymbol{S}}, t = \frac{\boldsymbol{S} \times \boldsymbol{U}.(\boldsymbol{X} - \boldsymbol{X}_0)}{\boldsymbol{S} \times \boldsymbol{U}.\boldsymbol{T}}, u = \frac{\boldsymbol{S} \times \boldsymbol{T}.(\boldsymbol{X} - \boldsymbol{X}_0)}{\boldsymbol{S} \times \boldsymbol{T}.\boldsymbol{U}} \tag{3.2}$$

If a point $\boldsymbol{X}$ lies in the grid then it satisfies $0 < s < 1, 0 < t < 1$ and $0 < u < 1$.

As a second step, the rectangular grid points have to be computed based on the size of the optical surface.The grid is cut into l, m, n parts such that its vertices form control points $\boldsymbol{P}_{ijk}$ containing $l + 1$ planes in the **S** direction, $m+1$ planes in the **T** direction and $n+1$ planes in the **U** direction. The locations of each control point in the grid is assigned using the equation

$$\boldsymbol{P}_{ijk} = \mathbf{X}_0 + \frac{i}{l}\mathbf{S} + \frac{j}{m}\boldsymbol{T} + \frac{k}{n}\boldsymbol{U} \tag{3.3}$$

where $i \in [0 \cdots l]$, $j \in [0 \cdots m]$, $k \in [0 \cdots n]$ and thus forming a rectangular lattice with $(l+1)(m+1)(n+1)$ control points.

The third step is the deformation step. The deformation is attained due to the displacement vector $\boldsymbol{\mu}_{ijk}$ acting along i or j or k or combinations of them shifting the grid point $\boldsymbol{P}_{ijk}$ to the new $\boldsymbol{P}'_{ijk}$ position.

$$\boldsymbol{P'}_{ijk} = \boldsymbol{P}_{ijk} + \boldsymbol{\mu}_{ijk} \tag{3.4}$$

As a last step, this deformation at the grid causes deformation of the underlying surface. Its new position $\boldsymbol{X}_{ffd}$ of an arbitrary point $\boldsymbol{X}$ corresponding to the deformation of the grid $\boldsymbol{P}'_{ijk}$ is found by computing its (s,t,u) coordinates from equation 3.2 and then evaluating the trivariate Bernstein polynomial [28] which is given by

$$\begin{aligned}\boldsymbol{X}_{ffd} = \sum_{i=0}^{l} \binom{l}{i} (1-s)^{l-i} s^i \\ \left[\sum_{j=0}^{m} \binom{m}{j} (1-t)^{m-j} t^j \left[\sum_{k=0}^{n} \binom{n}{k} (1-u)^{n-k} u^k \boldsymbol{P'}_{ijk}\right]\right]\end{aligned} \tag{3.5}$$

where $\boldsymbol{X}_{ffd}$ is a vector which contains the coordinates of the displaced point of the underlying surface.

For rotational symmetric systems, it is well enough to deform the freeform curves instead of the surfaces. In such cases, the freeform curves are placed inside a 2D grid and the FFD algorithm described above is applied. The deformation principle explained above hold for rotationally symmetric systems other than the dimension of the grid being reduced to two instead of three.

3.1.3 RAYTRACING

Once the optical surface is generated using deformation technique, optical performance has to be characterized. This is done using a standard raytracer.

After each raytrace, the optical performance is computed either as intensity or illuminance and these results help to evaluate the optical performance using the merit functions which will be described in the next section.

3.1.4 MERIT FUNCTION EVALUATION

Before doing an optimization, the optical designer has varying needs like maximizing the luminous flux at the target, homogeneity, appropriate shape of the light distribution, sharp gradient and so on. This varies depending on the application and the deployment of the designed optical system. The mathematical function which could represent these varying optical performances is called the merit function and the optimization algorithm needs to either maximize or minimize them. The evaluated merit function is represented using a single numerical value described as Q. This Q captures all aspects of the performance of an optical surface. When the modeled merit functions are accurate, the convergence will be faster if a solution exists.

In illumination design problems, it is often necessary to maximize the luminous flux in the target area and to maximize its uniformity by reducing the deviation between the actual and the prescribed lighting requirements. This type of photometric measures are mathematically converted in the OFFD as deviation based and luminous flux merit functions and are described as follows.

DEVIATION BASED MERIT FUNCTION

The deviation based merit function evaluates how far the actual distribution varies from the desired lighting requirements. In this method, the evaluated value represented as Q_{dev} is calculated by evaluating the deviation of each corresponding pixel (x) between the current and the desired distribution which can be expressed as

$$Q_{dev} = \int_G [E_{ideal}(x) - E(x)]^2 \, dx \tag{3.6}$$

where G is the target area one is interested in, $E_{ideal}(x)$ is the desired distribution and $E(x)$ is the current distribution.

LUMINOUS FLUX BASED MERIT FUNCTION

In contrary to the deviation based merit function, the luminous flux based merit function gives information about the amount of luminous flux redirected by the optics to the desired region. This does not carry any information regarding the shape or uniformity of the light distribution. To maximize the optical efficiency, this merit function is commonly used. The evaluated value represented as Q_{flux} is calculated by comparing the total luminous flux in the target area G of the current system with the total flux available from the light source. The Q_{flux} is calculated using

$$Q_{flux} = 1 - \frac{\Phi_{target}}{\Phi_{source}} \tag{3.7}$$

Based on the requirements than an optical designer is aiming for, the corresponding suitable merit function will be selected. This appropriate selection speeds up the optimization process.

3.1.5 OPTIMIZATION

An optimization algorithm is a guidance tool for problems when there is no straight forward solution and when the search space is large. In the freeform deformation technique, an optimal surface is created by deforming the rectangular grid. There exist infinite possible ways to deform this grid and infinite surfaces could be created. Then all theses surfaces have to be analyzed one after the other using a raytracer and evaluate them. But the complexity of analyzing each of the deformed surface is enormous and it takes more time to finish this task and this approach even could end up without having the best possible surface. In contrary, by using a suitable optimization algorithm with rectangular grid points $\boldsymbol{P}_{ijk}$ as optimization variables, an optimal surface can be generated. The optimization algorithm starts with an initial guess for its optimization variables and then based on the figure of merit, they generate a sequence of estimates iteratively until they reach the best solution [30]. So optimization plays a huge role in finding the best possible displacements of the grid control points which is then translated into a best optical surface. OFFD employs Nelder-Mead simplex technique as its optimization algorithm because it is derivative free and stable.

PROPOSED AREAS FOR IMPROVEMENT

To direct the optimization in the correct direction and to obtain surfaces suitable to manufacture, surface analysis must be made during the routine. Further, advanced optimization algorithms must be implemented for faster convergence. Different surface representations can also be a part of the system when sharp and local deformation is needed. To improve the speed and to make it more autonomous, optimal grid points must be selected. These aspects are explored as a part of this dissertation and they are covered in detail in the coming chapters using the below-described examples.

3.2 DESIGN EXAMPLES IN THE DISSERTATION

STREET LIGHTING LENS

As a first example, an optical design for the street lighting system used in the work of Wendel [4] is taken. The setup of this task is as shown in the figure 3.4. A Cree XPG2 LED [31] with 100 lumens is used as a light source and the initial optics comprises of a spherical surface that will be subjected to optimization and a plain end surface. The spherical lens before optimization is shown in figure 3.5(a).

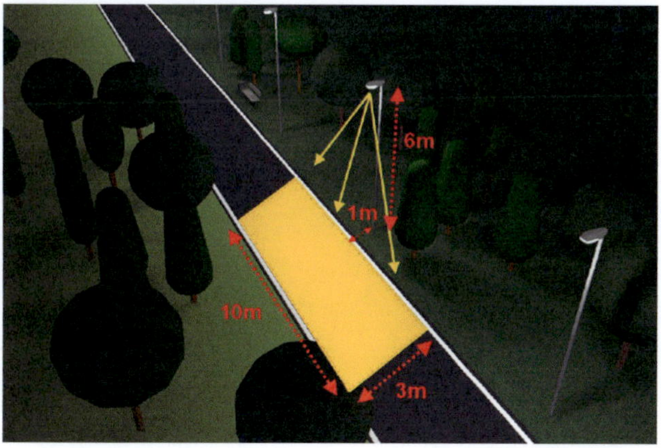

Figure 3.4: Schematic view of the street light setup with 10m pole spacing, 6m height, 1m away from the road. The yellow rectangle shows the area to be illuminated [4].

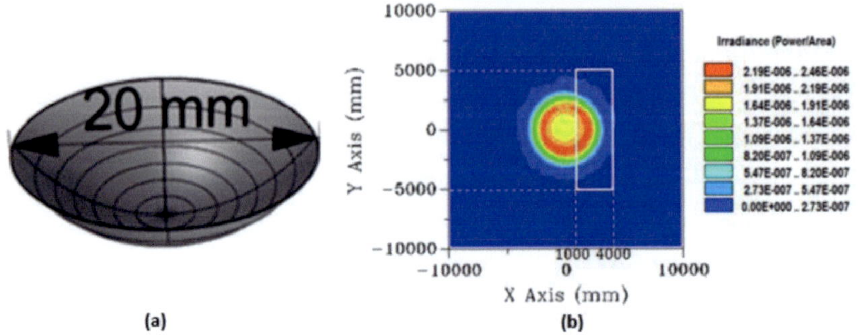

Figure 3.5: (a) initial optical surface used for optimization of the street lighting system and (b) the illuminance distribution created by the initial spherical surface with the white rectangle showing the target distribution.

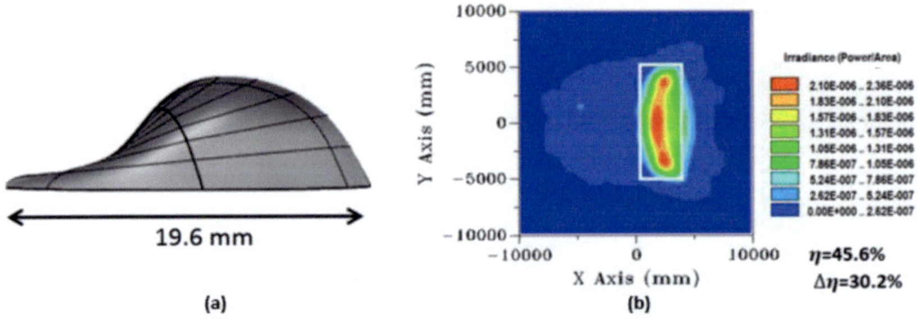

Figure 3.6: (a) deformed and optimized optical surface and (b) the illuminance distribution using the previous work [4].

The efficiency of the initial surface η_{ini}, is only 15.4%. Moreover, the shape of the initial distribution is far away from the needed distribution (marked as a white rectangle) in the figure 3.5(b). The initial surface is optimized using the OFFD technique and the final results based on the previous work [4] yields an optical efficiency (η) of 45.6%

after 1500 iterations. Different combinations of grid points were used as optimization variables and they are evaluated using deviation and flux based merit functions alternatively. So the efficiency improvement given by $\Delta\eta = \eta - \eta_{ini}$ is 30.2%. The final optical surface and the light distribution is shown in the figure 3.6.

STOP LAMPS FOR AUTOMOTIVE LIGHTING

As a second example to show the potentiality of this work, an automotive stop lamp has been taken. The stop lamp is designed using TIR hybrid optics which contains a single refractive surface, total internal reflecting (TIR) surface, side entry surface which will be subjected to deformation and plain surface as an end plane which is shown in the figure 3.7. The designed stop lamp needs to fulfill the light intensity distribution specified by the United Nations Economic Commission for Europe (UNECE) and these requirements are given in the table 3.1.

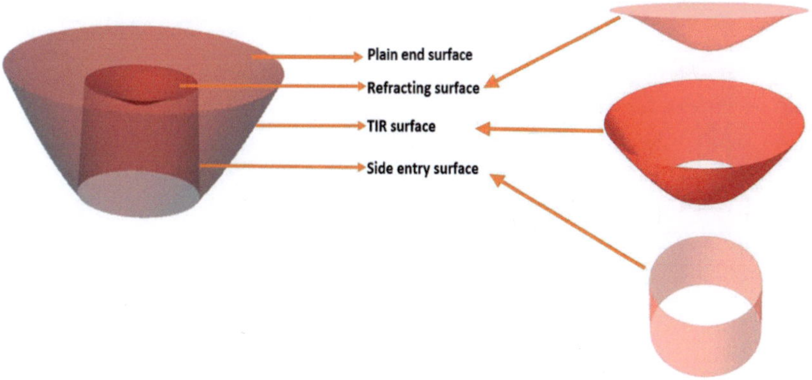

Figure 3.7: Representation of different surfaces in the hybrid TIR optics for automotive stop lamp.

The values specified by the UNECE are then interpolated to create a target distribution as shown in the figure 3.8.

[deg]	-45°	-30°	-20°	-10°	-5°	0°	5°	10°	20°	30°	45°
15°	0.3	0.3	0.3	0.3	0.3	0.3	0.3	0.3	0.3	0.3	0.3
10°	0.3	-	-	-	12	12	12	-	-	-	0.3
5°	0.3	-	6	12	-	42	-	12	6	-	0.3
0°	0.3	-	-	21	54	60	54	21	-	-	0.3
-5°	0.3	-	6	12	-	42	-	12	6	-	0.3
-10°	0.3	-	-	-	12	12	12	-	-	-	0.3
-15°	0.3	0.3	0.3	0.3	0.3	0.3	0.3	0.3	0.3	0.3	0.3

Table 3.1: Intensity requirements for the automotive stop lamp, UNECE R7 S1 [32].

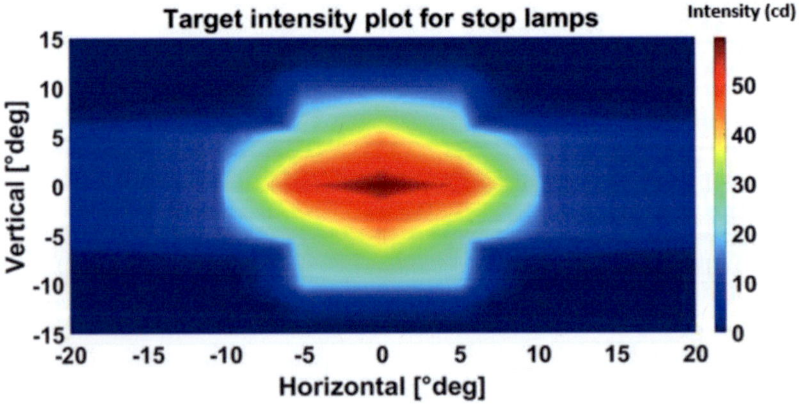

Figure 3.8: Target intensity distribution for stop lamp based on UNECE R7 S1 interpolated from the table 3.1.

A Luxeon Rebel LED from Lumileds [33] with 20 lumens is taken as a light source and the selected initial TIR hybrid optics is 28 mm in diameter and 11 mm thickness as shown in the figure 3.9. The simulated intensity distribution for this initial TIR hybrid optics is shown in the figure 3.10.

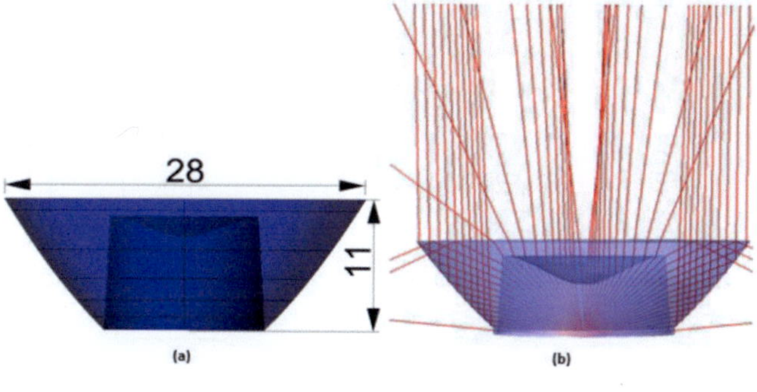

Figure 3.9: (a) Initial TIR hybrid optics used in the optimization of a stoplamp with all dimensions in millimeters and (b) schematic view of its rayfan.

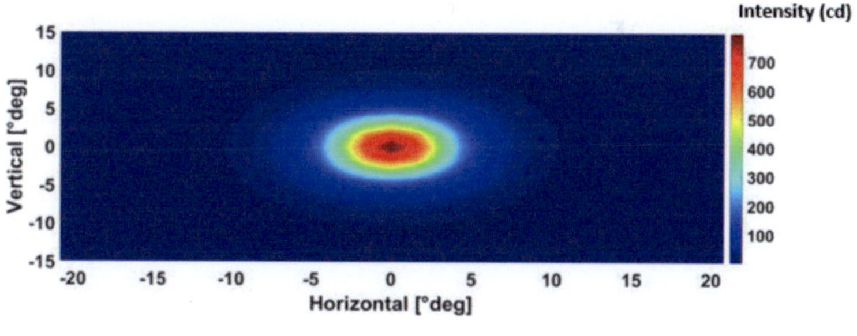

Figure 3.10: Intensity distribution obtained from the initial TIR hybrid optics.

From the intensity distribution, it can be seen that it is collimated with high intensities around -2° till +2° and 810 cd in the main beam direction (0°). This means that this distribution fails to satisfy the legal requirements because of its high intensity at the middle and less flux elsewhere as seen in the table 3.2. It is also to be noted that 95% of the collected luminous flux lies inside the needed region as opposed to the street lighting lens. So optimization is not required to maximize the flux in the targeted region but the extra luminous flux in the middle must be optimally distributed based on the legal requirements.

[*deg*]	-45°	-30°	-20°	-10°	-5°	0°	5°	10°	20°	30°	45°
15°	0.08	0.05	1	3	4	4	4	3	1	0.05	0.08
10°	0.1	-	-	-	9.5	14	9.5	-	-	-	0.1
5°	0.1	-	3	9	-	156	-	9	3	-	0.1
0°	0.1	-	-	15	156	810	156	15	-	-	0.1
-5°	0.1	-	3	9	-	156	-	9	3	-	0.1
-10°	0.1	-	-	-	9.5	14	9.5	-	-	-	0.1
-15°	0.08	0.05	1	3	4	4	4	3	1	0.05	0.08

Table 3.2: Luminous intensity at test points obtained from the initial TIR hybrid optics with red indicating the failed ones.

These two varied design examples are selected to verify the implemented techniques into the OFFD which will be discussed in the coming chapters.

CHAPTER 4

EXPLORATION OF OPTIMIZATION ALGORITHMS

The time intensive component in the OFFD is the raytracing of optical surfaces. Therefore, a fast converging and a more robust optimization system is necessary for making the OFFD an effective tool. A good optimization algorithm yields an improvement in the light distribution at a faster rate and this chapter explores different optimization algorithms in a way to find an appropriate one for the OFFD.

4.1 BACKGROUND

The optimization in the OFFD is implemented using the Nelder-Mead simplex algorithm in the previous work [4]. The simplex algorithm is commonly used in most of the illumination design problems because it is derivative free and stable [34]. In the recent years, several authors have proposed many variants of the Nelder-Mead simplex algorithm. Moreover, the popularity of genetic algorithm is also increasing. So it is a natural choice to explore and implement these algorithms in the OFFD and analyze its performance. This chapter shows how these algorithms are implemented in OFFD, their comparison results, and insights drawn as a result of this exploration. The algorithms are validated using the street lighting system introduced in the section 3.2

of the previous chapter. The initial system, lighting requirements and merit functions remain the same.

GLOBAL AND LOCAL OPTIMIZATION

The optimization can be classified as a global and local optimization. They are implemented to find either a global or a local minimum point. The local minimum point is the location at which the evaluated value of the merit function is smaller than all the other adjacent points, but not necessarily in the whole domain [30]. In contrary, the global minimum is the location at which the merit function evaluates to a value which is the minimum among all the points in the whole domain of the function. Nelder-Mead simplex is one such example for local search while genetic algorithm is an example for global search optimization. These two algorithms are explored fully in the next sections.

4.2 NELDER-MEAD SIMPLEX ALGORITHM

The Nelder-Mead simplex otherwise called as downhill algorithm is a direct and local search optimization method [35]. It uses geometric approach to achieve minimization [36] by constructing an $n-$ dimensional, closed geometric convex hull using $(n+1)$ vertices known as simplex. The function is evaluated at each vertex and the vertex with the highest value (worst vertex) is iteratively replaced by another vertex [35]. The algorithm thus moves down to reach the optimum [37]. This downward movement of simplex algorithm is done with the help of four operations. They are reflection, contraction, expansion and shrink and the corresponding operators are notated as α, β, γ and

ρ respectively. The algorithm is explained in the appendix A where the meaning of each operator is elaborated in detail. Due to its extensive use of the Nelder-Mead simplex algorithm for various applications, several authors have studied it and tweaked its parameters to make the convergence faster. The values of the operators based on their findings are described in the coming section. In the next section, these parameters are applied to the OFFD and the optical performance will be compared against each other using the street lighting lens as a design example.

4.2.1 SIMPLEX VARIANTS

DEFAULT NELDER-MEAD SIMPLEX METHOD (DNMS)

The operators of default Nelder-Mead as specified by Nelder and Mead [35] must satisfy $\alpha > 0, \gamma > 1, 0 < \beta < 1, 0 < \rho < 1$. The values of these operators are given in the equation 4.1.

$$\{\alpha, \gamma, \beta, \rho\} = \{1, 2, 0.5, 0.5\} \tag{4.1}$$

ADAPTIVE NELDER-MEAD SIMPLEX METHOD (ANMS)

Gao and Han found that the expansion and contraction in the simplex begin to decrease when the objective function is convex [38]. They studied this inefficient behavior for a large number of optimization variables and found that the operators are influenced by the size of the optimization variables. It means the operators are not any more constants as defined by Nelder and Mead. They found a new combination based on the total number of optimization variables (d) and it is given in the equation 4.2.

$$\{\alpha,\gamma,\beta,\rho\} = \left\{1, 1+\frac{2}{d}, 0.75-\frac{1}{2d}, 1-\frac{1}{d}\right\} \tag{4.2}$$

For example, when five control points are selected from the OFFD grid shown in the figure 4.1, the total number of optimization variables (d) when applied to a 3-dimensional system will be $d = 3 \times 5 = 15$.

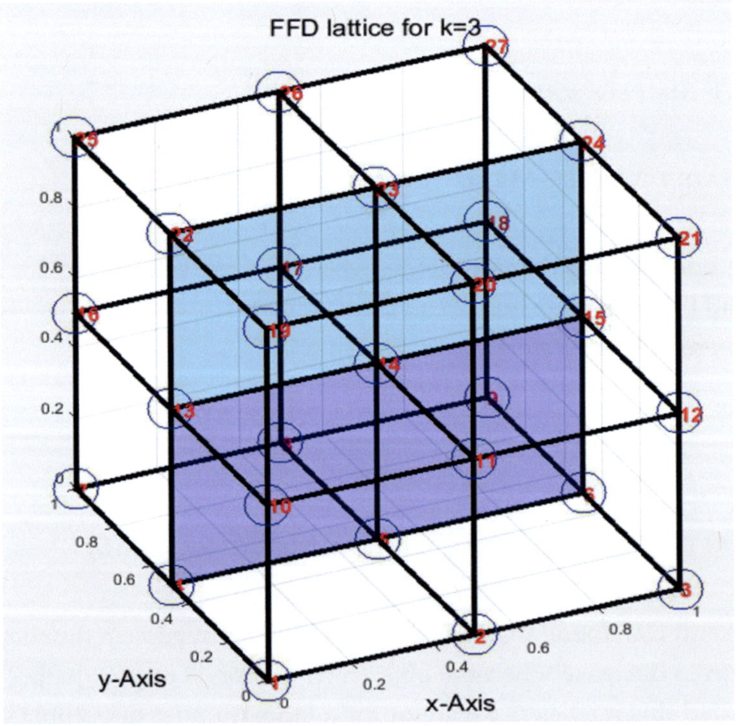

Figure 4.1: OFFD grid showing the control grid points that are numbered inside the blue circles for identification. These control grid points will be used as optimization variables by the OFFD.

KOSHEL'S METHOD

R. J. Koshel proposed new operator values for the Nelder-Mead simplex by investigating the shape of the figure of merit space. He found that the figure of merit space is parabolic near the local minima. So he optimized them for a total number of optimization parameters ranging from 2 to 15. He claims that his new set of values showed improvement in the optical performance and faster convergence [36]. Table 4.1 shows the parameters values of all the operators suggested by him for a specific number of optimization variables (d).

d	α	γ	β	ρ
3	1.0047228	2.01219185	0.40248282	0.63582727
6	1.01666875	1.49081084	0.52542045	0.63782812
9	1.05786796	1.44480667	0.61070155	0.9669131
12	1.07848799	1.40881131	0.64421344	0.53868179
15	0.76135727	1.35150713	0.67405069	0.09221549

Table 4.1: Operator values for the Simplex algorithm suggested by R.J.Koshel [36].

S9 METHOD

The S9 method as proposed by Barton and Ivey studied the behavior of simplex method by optimizing the expected response of a stochastic system. When the default simplex operators were used, it led to inappropriate termination [37]. Based on their analytical and empirical results, the shrink step was altered by increasing the shrink operator ρ

from 0.5 to 0.9, thereby reducing the simplex geometry by only 10% rather than 50% [37]. The values are stated in the equation 4.3.

$$\{\alpha, \gamma, \beta, \rho\} = \{1, 2, 0.5, 0.9\} \tag{4.3}$$

FAN AND ZAHARA METHOD

Fan and Zahara investigated the performance of Nelder-Mead operators for several functions like Powell function, Beale function, Helical valley function, Box three dimensional functions and Wood function [39]. They selected the best parameter values based on the highest percentage rate on successful minimization of all these functions. The suggested set of operator values are given in equation 4.4.

$$\{\alpha, \gamma, \beta, \rho\} = \{1.5, 2.75, 0.75, 0.5\} \tag{4.4}$$

WANG AND SHOUP METHOD

Fan and Zahara' s study was further followed by an extended and comprehensive parameter sensitivity research by Wang and Shoup. They used seven different test functions like B2, Beale, Booth, Wood, Rastrigin, Rosenbrock and sphere functions to search for any common patterns and the relationship of each parameter in producing the optimum solution [40]. Based on their analysis, the results are concluded in the equation 4.5.

$$\{\alpha, \gamma, \beta, \rho\} = \{1.29, 2.29, 0.47, 0.57\} \tag{4.5}$$

4.2.2 IMPLEMENTATION OF SIMPLEX VARIANTS IN OFFD

Though several researchers have studied the operator values of the simplex extensively, it still remains unknown whether their suggested values brings improvement to the OFFD. Therefore these simplex variants must be implemented in the OFFD and its performance have to be analyzed which is covered in this section. The street setup, initial spherical surface, the lighting requirements and merit functions to evaluate the photometric performance described in the previous chapter are used. The control grid points are selected based on the OFFD grid shown in the figure 4.1. To understand the influence of optimization variables in the optical performance, different control grid combinations are used to analyze all the simplex variants. The optimization is carried out using deviation based merit function for 300 iterations and then the optical performance of the optimized surface (Q_{dev}) is evaluated. The best result is again optimized using the flux based merit function for 300 iterations and the optical efficiency η of the optimized surface is found out. The selection order of the merit function is just arbitrary and even the reverse order does not bring any difference in the performance of the system as the same initial surface is used. But optimization using both merit functions is necessary as the street lighting design task require maximum flux in the targeted rectangular region as well as high uniformity in the desired region. The improvement in Q_{dev} and η are denoted as ΔQ_{dev} and $\Delta\eta$ and are defined in equations 4.6 and 4.7 respectively.

$$\Delta Q_{dev} = \frac{|Q_{dev} - Q_{ini}|}{Q_{ini}} \tag{4.6}$$

where Q_{ini} is the numerical value specifying the quality of the initial optical surface.

$$\Delta\eta = \eta - \eta_{ini} \tag{4.7}$$

where η_{ini} is the optical efficiency of the initial optical surface. The results of ΔQ_{dev} after 300 runs for all the simplex variants are shown in the table 4.2. The obtained surfaces are optimized again using the flux merit function for 300 iterations to improve the luminous flux in the needed area. The results of $\Delta\eta$ are furnished in the table 4.3.

variants	9	12	18
DNMS	83.4	**86.3**	73.1
ANMS	**86.8**	81.6	72.7
Koshel	**86.8**	82.4	NA
S9	83.4	**86.3**	**83.9**
Fan-Zahra	82.1	68.4	70.5
Wang-Shoup	79.1	69.4	82.8
Average	83.6	79.1	76.6

Table 4.2: Improvement in the shape of the light distribution ΔQ_{dev} after 300 iterations for different number of optimization variables like $9, 12, 18$ using all Nelder-Mead variants and the underlined results are the top performers for that particular number of optimization variables. Koshel's method did not provide any suggestion when the total number of optimization variables equals 18.

The optimized illuminance distributions of the street lighting lens after 600 iterations using the simplex variants are shown in figures 4.2, 4.3 and 4.4 for a number of optimization variables as 9, 12, 18 respectively. When the total number of optimization variables is selected as nine, no distinct difference is seen in the shape of the light distribution between simplex variants as in the figure 4.2. But the luminous flux in the rectangular region differs based on the simplex variant. Out of all the variants, Koshel's method and ANMS method performs well compared to others. When the total number of optimization variables are 12 and 18, then there is a significant difference between the simplex variants as seen in figures 4.3 and 4.4. When the total number of optimization variables is 12, the DNMS and S9 method perform well with an improvement in the efficiency of about 40%. But when increasing the number of optimization variables to 18, the performance diminishes and only S9 performs relatively well.

variants	9	12	18
DNMS	21.7	**39.8**	22.2
ANMS	33.5	34.5	21.9
Koshel	**34.1**	34.3	NA
S9	21.7	**39.8**	**32.4**
Fan-Zahra	31.1	20.5	26.5
Wang-Shoup	22.2	22.0	30.0
Average	27.4	31.8	26.6

Table 4.3: Improvement in the luminous flux in the targeted region $\Delta\eta$ after 600 iterations for different number of optimization variables like $9, 12, 18$ using all Nelder-Mead variants and the underlined results are the top performers for that particular number of optimization variables. Koshel's method did not provide any suggestion when the total number of optimization variables equals 18.

INFERENCES

For the OFFD, the number of optimization variables used in the deformation system plays a significant role in the performance. Hence different variants of the simplex are analyzed by varying the number of optimization variables. After the analysis, it is found out, when the optical designer selects nine optimization variables, it is recommended to use Koshel's method or ANMS. When he selects 12-15 optimization variables, he is free to use S9 or DNMS methods and when the number of optimization variables is increased further, the performance depreciates. Based on this study, overall it is highly recommended to use 12-15 optimization variables for better and faster results. As simplex is a local search based algorithm, this exploration is continued further with the global optimization in the next section.

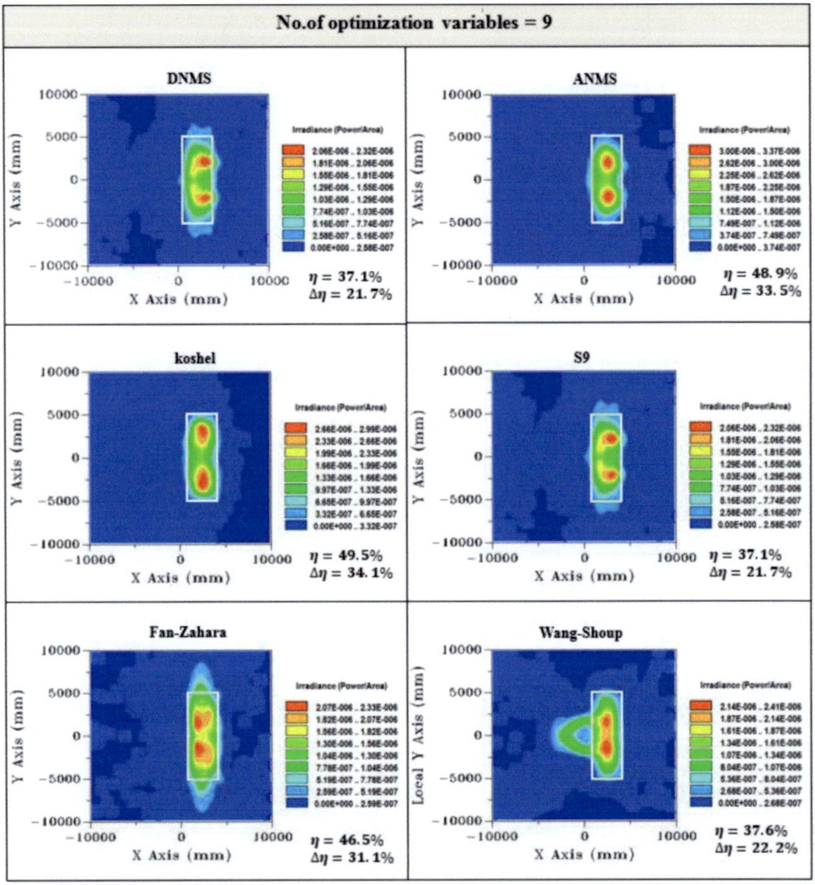

Figure 4.2: Simulated light distribution on street area (marked in white) after 600 iterations for the simplex variants using 9 optimization variables.

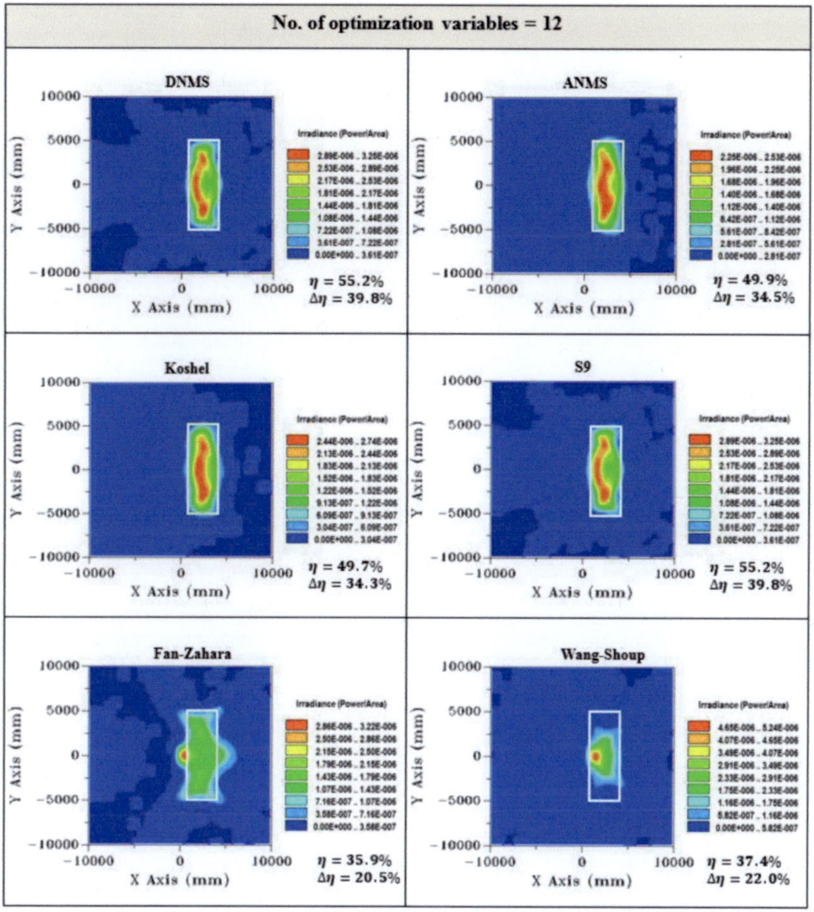

Figure 4.3: Illuminance distribution on street (marked in white) after 600 iterations for the simplex variants using 12 optimization variables.

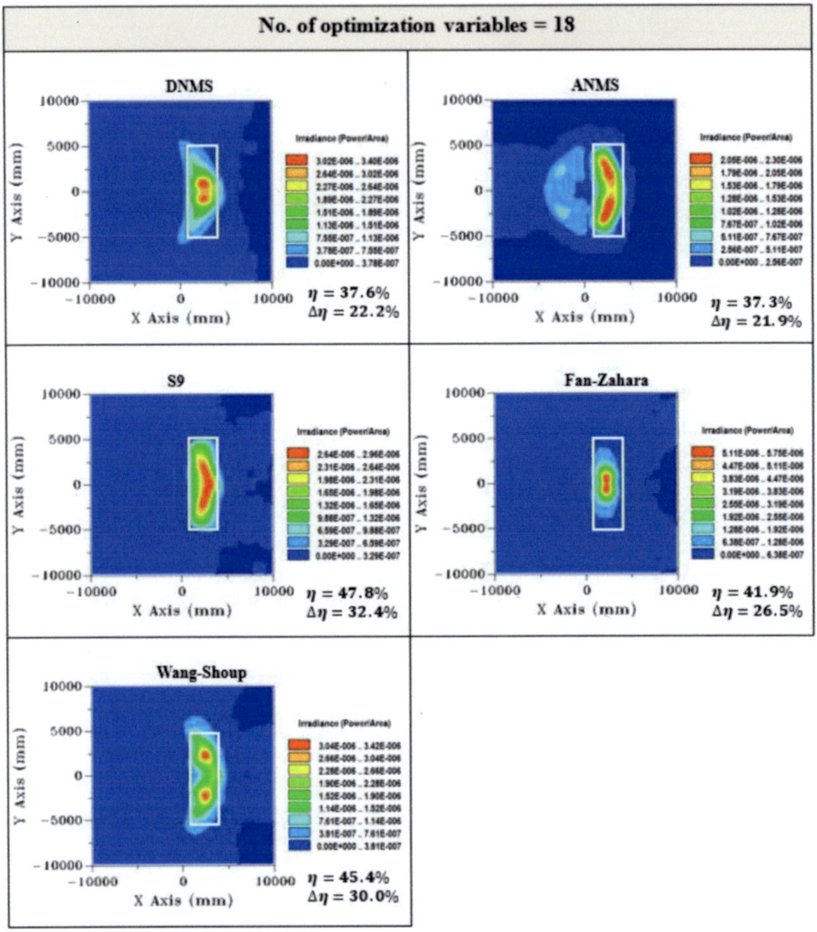

Figure 4.4: Illuminance distribution on street (marked in white) after 600 iterations for the simplex variants using 18 optimization variables.

4.3 GENETIC ALGORITHM

The main drawback of the simplex is that the vertices computed in each step are based only on the previous iteration. Hence, Nelder-Mead simplex method gets stuck in its local minima easily [41, 42]. In order to find global minima by searching in a large landscape, a global search algorithm is necessary. As genetic algorithm is used widely in the computational optimization and operations research due to its operational simplicity, it is a good choice to study and implement it in the OFFD. The theory, implementation of this algorithm along with the parameter study and evaluation results using the street lighting lens are detailed in this section.

4.3.1 THEORY

The genetic algorithm is inspired by Darwin's theory of evolution, the phenomenon seen quite in nature. According to it, individuals develop or evolve naturally in accordance with the environment by showing an improvement in every generation [43]. In the same manner, the genetic algorithm starts with a set of an initial population called chromosomes that are randomly selected [44]. The whole processes involved in the genetic algorithm is depicted using the flowchart in figure 4.5.

In order to mimic "survival of fittest" of nature's selection mechanism, the evaluated merit function values of initial population are calculated and ranked. The process of evolution, i.e. generation of new chromosomes, is then followed. The evolution process is achieved using several operations like selection, reproduction and mutation. They select the chromosomes to be inherited in the next generation by eliminating the inferior ones. And also, these operations define how the genomes exchange with each other and move forward.

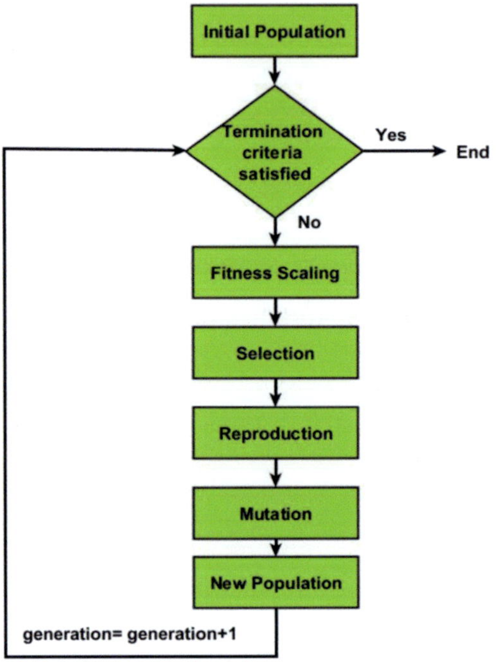

Figure 4.5: Flow chart of a genetic algorithm [41].

4.3.2 Implementation of Genetic Algorithm in OFFD

Because of the huge involvement of various parameters in each step, the genetic algorithm performs differently for each distinct problem [45]. The interaction between these factors and its influence in the results has been studied by many researchers. However, there is no generalized solution so far on how to select these parameters. As the operational methods and their parameters hold no direct relationship with OFFD, specific parameter study is necessary in order to find the best possible solution in a short time. So this section explains the parameter sensitivity analysis that was carried out to analyze the parameters of the genetic algorithm to make it suitable for the OFFD.

No.of optimization variables (d)	Grid point combinations (GP)
9	[1,3,11], [2,4,6]
12	[1,3,13,15], [1,3,10,12]
15	[1,3,11,13,15], [2,4,6,10,12]

Table 4.4: Description of the test cases for the parameter study in genetic algorithm.

A study was conducted first by varying the number of optimization variables (d). Different combinations of control grid points (GP) were taken into account and the selected combinations of grid points are illustrated in the table 4.4. The parameter study for these operators was carried out for the street lighting lens using the deviation based merit function.

INITIAL POPULATION

Genetic algorithm starts with the initialization of the optimization variables called as chromosomes. In our case, the shifts of the control grid points are the chromosomes referred as

$$chromosomes = [x_1, x_2, \ldots, x_d] \tag{4.8}$$

The merit function is then evaluated on this grid of control points to find a minimum value commonly known as Q given by

$$f\,(chromosome) = f\,(x_1, x_{2,\cdots}, x_d) = Q \tag{4.9}$$

Initially, the optimization algorithm begins by generating random shifts for the selected grid points known as population. For a population size z and number of control grid point shifts as d, the initial population will be generated as $d \times z$ matrix. For a larger population size, the algorithm could search on a wider space and therefore it can achieve a more global result. But at the same time, large population size slows down the convergence speed, which makes the selection of population size a little tricky. So a study was carried out for different population sizes and its optical performance (Q) was plotted in the graph as shown in the figure 4.6. From the results, it is obvious that the OFFD performs better when the initial population size is around 10-20 and deteriorates as the population size grows. This is because, as the population size increases, the randomness in the optimization process gets increased which leads to a slower convergence.

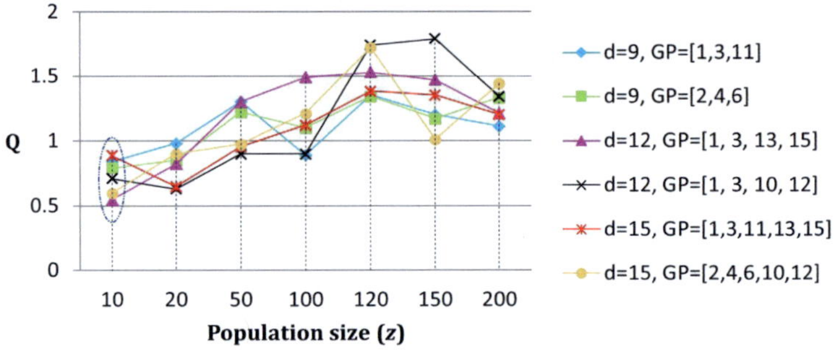

Figure 4.6: Graphical representation showing the performance of Q for varying population size (z) for different control grid point combinations.

FITNESS SCALING

In nature, only the creatures which own high fitness than the surrounding environment can evolve and unfit species are discarded. The raw fitness scores of different optimization variables are mathematically calculated using the deviation and the flux based merit functions.

$$\begin{aligned}
(raw\ score)_1 &= f\,(chromosome_1) &= f\,(x_{11}, x_{12}, \ldots, x_{1z}) \\
(raw\ score)_2 &= f\,(chromosome_2) &= f\,(x_{21}, x_{22}, \ldots, x_{2z}) \\
\vdots \quad & \qquad \vdots & \vdots \\
(raw\ score)_n &= f\,(chromosome_d) &= f\,(x_{d1}, x_{d2}, \ldots x_{dz})
\end{aligned} \tag{4.10}$$

Scaling based on rank is used as it is more appropriate because lesser the Q, better the optical performance and more chance to participate in the next iteration. In this step, raw scores from $1, 2, \ldots, n$ are sorted and a rank (r) is assigned based on its performance.

SELECTION

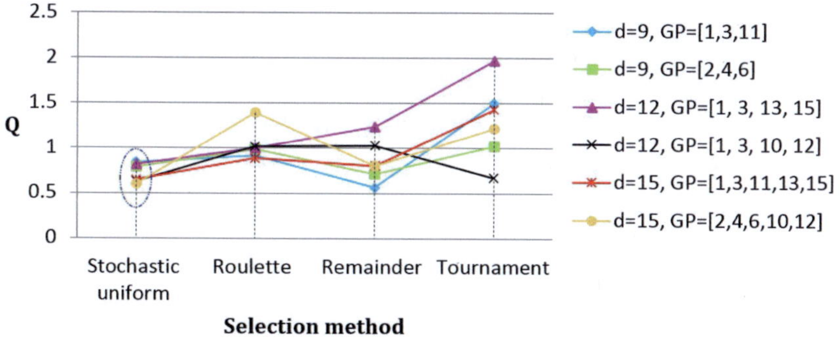

Figure 4.7: Graphical representation showing the performance of Q for different selection methods in OFFD.

The genetic algorithm uses an operator to select the good performers to proceed with the next generation. The four operators based on several research studies are stochastic uniform, roulette, remainder and tournament [46]. These four functions are evaluated in the OFFD to select the best individuals and the results are shown in the figure 4.7. Irrespective of the control grid points, the remainder method and stochastic uniform method outperforms others. But as the size of the dimension (d) increases, the remainder method could not perform well compared to its counterpart thereby making stochastic uniform function as the best selection method for the OFFD.

REPRODUCTION

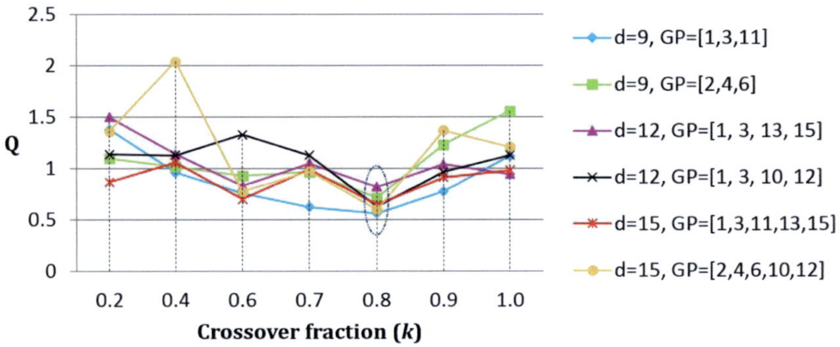

Figure 4.8: Graphical representation showing the performance of Q for various crossover fractions in OFFD.

The optimization variables selected in the last step are used to reproduce better individuals in this step. Elitism and crossover are the two main operations involved here. The elites are those individuals who remain unchanged when participating in the next generation. Based on the statistical analysis, 5% of the total population size is usually selected as elites. The crossover fraction (k) defines the fraction of optimization variables selected for the next generation. This can be generated using the operators like a single point, multi-point, scattered and heuristics crossover. So different crossover fractions and various crossover functions are tried in OFFD and the performance is shown in figures 4.8 and 4.9. OFFD performs well for a crossover fraction of $k = 0.8$ and scattered crossover function.

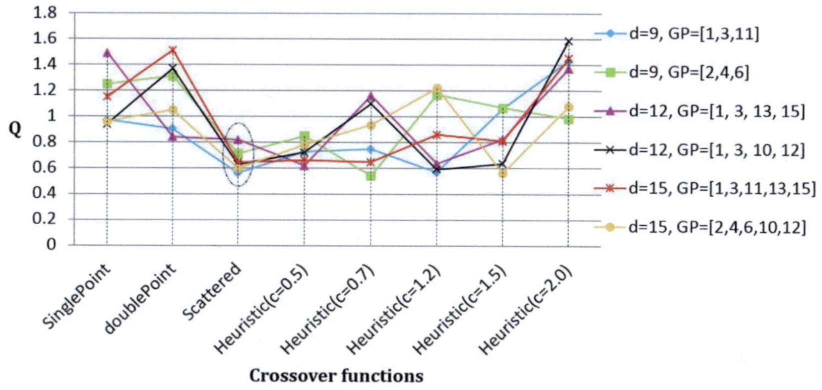

Figure 4.9: Graphical representation showing the performance of Q for different crossover functions in OFFD.

MUTATION

In order to maintain genetic diversity, mutation is applied after the reproduction process. During this step, the genetic algorithm selects variables from a large search space by randomly and slightly changing its values. This leads to a higher probability of finding the global minimum. This mutation process can be realized using statistical methods like uniform and Gaussian. The chance of being mutated is specified as the mutation rate (t) in the uniform distribution and as variance (σ) in the Gaussian distribution. The OFFD is guaranteed to perform well when the Gaussian function with variance $(\sigma = 1)$ is selected as shown in the figure 4.10. Thus the parameter sensitivity analysis for each operator in the genetic algorithm was carried out. This included optimal number of optimization variables to be selected as an initial population, the nature's phenomena being modeled using various mathematical definitions and its compatibility with the OFFD. These results are summarized in the table 4.5 and it is highly recommended to use them to get better optical performance when genetic algorithm is used as a optimization technique in the OFFD.

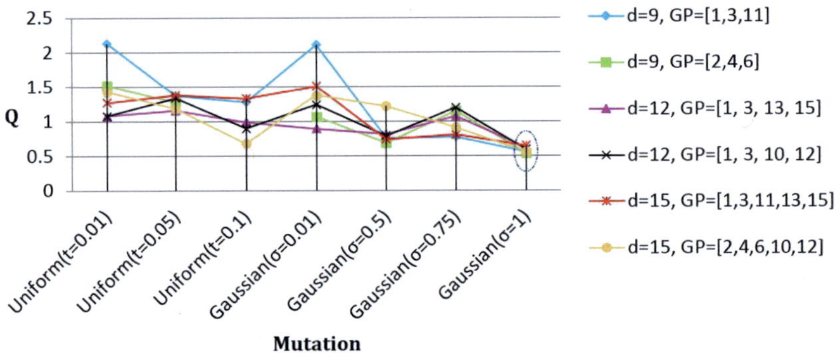

Figure 4.10: Graphical representation showing the performance of Q for various statistical functions for the mutation operation.

VALIDATION

The optical performance has to be analyzed based on the operators shown in the table 4.5. Just like the Simplex, the optimization is performed using deviation based merit function for 300 iterations followed by a flux based function for 300 iterations. For simplicity, the control grid points selected to analyze its performance are $[2, 4, 6]$ and $[1, 3, 13, 15]$. After 600 iterations, the optimized result with a significant improvement in efficiency was obtained and is shown in the figure 4.11. The optimized surface yielding uniform illumination in the target region is shown in the figure 4.12.

Operator	Recommended values /functions
Number of generations	300
Population size (z)	10-20
Fitness scaling	Rank
No.of elites	5% of the population size
Selection method	Stochastic uniform
Crossover fraction (k)	0.8
Crossover	Scattered crossover
Mutation	Gaussian with variance ($\sigma = 1$)

Table 4.5: Summary of the parameters of the genetic algorithm recommended for the OFFD based on the sensitivity analysis.

4.4 COMPARISON OF ALGORITHMS

The main difference between the simplex and the genetic algorithm is its approach in searching the best possible result. To search for a global minimum, randomness is introduced in every step of the optimization in genetic algorithm whereas the simplex depends solely on the previous iteration. From OFFD perspective, the choice of either simplex or genetic algorithm depends on the initial system. If the performance of the initial system is far away from the target, then a global search is essential to improve its performance significantly. Whereas, if the initial system requires a slight change, then a local search using simplex is faster and sufficient. The influence of the number of optimization variables is more pronounced in the simplex which is evident in the figures 4.2 - 4.4. Therefore an appropriate simplex variant has to be selected based on the number of optimization variables. But genetic algorithm is not sensitive to its underlying optimization variables and the optical designer is free to select the number of optimization variables based on his needs.

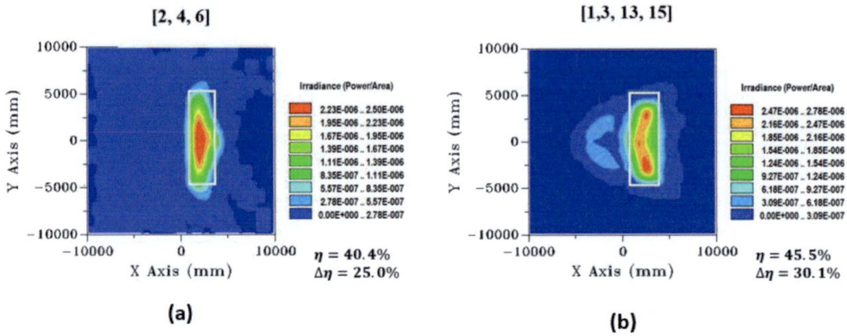

Figure 4.11: Illuminance distribution after 600 iterations using genetic algorithm (a) $[2,4,6]$ and (b) $[1,3,13,15]$ as optimization variables.

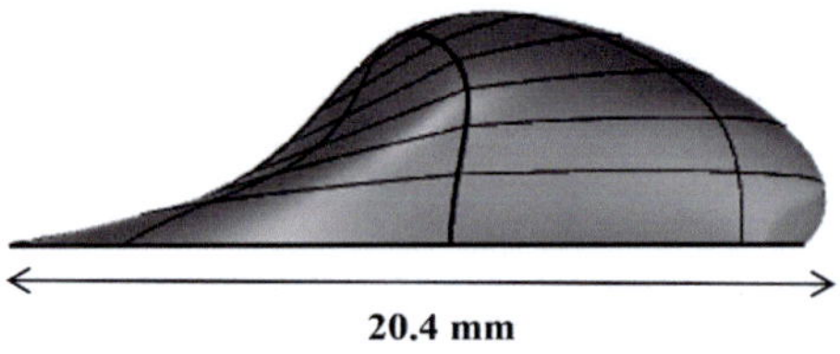

Figure 4.12: Deformed surface after 600 iterations using the genetic algorithm.

4.5 CONCLUSION

This chapter showed the exploration of Nelder-Mead simplex variants and genetic algorithm. The influence of the optimization variables towards OFFD and the parameters present in each algorithm is extensively studied and compared against each other. Each of the algorithms has its own advantages and drawbacks. But this study has provided some insights to select a suitable variant based on the design needs.

In the previous work, the improvement in optical efficiency $\Delta\eta$ is about 30.2% after 1500 iterations. By properly choosing the simplex variant for the selected number of optimization variables, one could attain the same improvement in just 600 iterations. But this was attained without the consideration of the manufacturing feasibility analysis. This major issue is addressed in the chapter 6.

Another big challenge is the selection of the appropriate control grid points. The grid points selected to show the performance of genetic and simplex variants in the chapter are the ideal ones. They are selected based on the prior knowledge of the optical designer. This drawback is addressed and an appropriate solution is provided in the chapter 7.

Throughout the optimization, the optical surface is represented using NURBS which is good in performing global deformation but cannot offer local modification of the optical surface. This can be solved by representing the optical surface using an alternate form which is elaborated in the coming chapter.

CHAPTER 5

LOCAL DEFORMATION

While global freeform deformation could be achieved using NURBS, there are certain lighting requirements where local deformation is necessary for attaining better results. This chapter shows how local deformation could be achieved using T-Splines, implementation of T-Splines in the OFFD and ends with a comparison between NURBS and T-Splines.

5.1 LIMITATIONS IN NURBS

NURBS is a common mathematical form to represent conics, quadrics and freeform curves and surfaces [47]. NURBS techniques are so mature and they are used in computer-aided graphics systems as well as in raytracers. Due to its flexibility, the surfaces can be easily manipulated or modified by changing the control points or its weights during the optimization. NURBS surface is the parametric tensor product surface and can be defined using the equation 1.1.

OFFD using NURBS could attain global deformation which makes the manufacturing easier as it is devoid of discontinuities. But there are cases where a sharp gradient in the light distribution is necessary or the path of the light has to be changed drastically. In such situations, a slight local deformation brings significant improvement. If such tasks are carried using NURBS based OFFD, the optical surface must be deformed using a single control grid point. But during the routine, the surface gets pulled in or pushed out leading to surface irregularities as shown in the figure 5.1. Therefore, selecting a single control grid point and deforming a surface is not a good choice as it will not provide the expected local deformation. So it is clear that modifying the OFFD grid could bring no improvement.

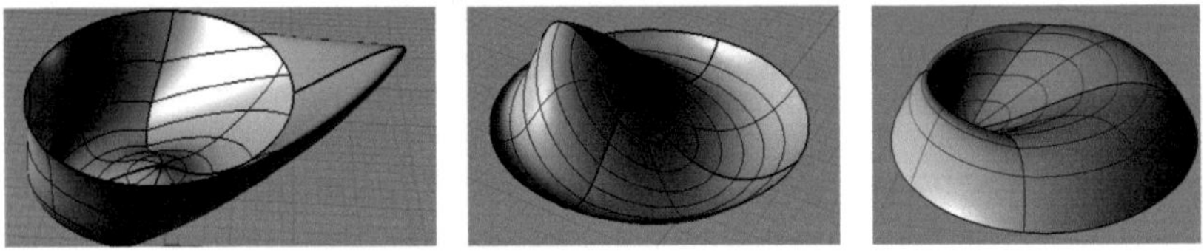

Figure 5.1: Samples of deformed optical surfaces that got pulled in or pushed out when a single grid point is selected as an optimization variable for the street lighting lens.

Another limitation in NURBS is that the NURBS model requires significant number of control points for its accurate representation.

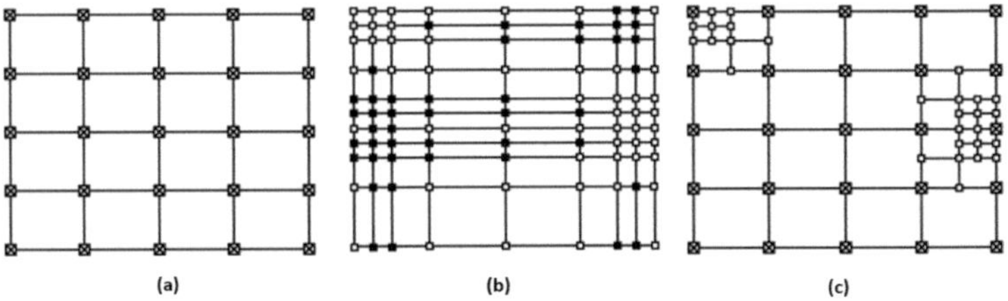

Figure 5.2: An example showing (a) a 5×5 NURBS patch and how the modification causes addition of control points (b) along rows and columns using NURBS and (c) only locally using T-Splines.

Furthermore, all the control points must lie in a rectangular grid which means that many control points are present merely to satisfy the topological constraints and they carry no information [48]. As seen in the figure 5.2, knot intervals of NURBS surface are repeated horizontally column by column and row by row vertically [49]. In order to satisfy this balance, if one adds a single knot vector, the entire column or row of control points gets simultaneously added [50]. This makes local refinement very hard to achieve [48].

The knot insertion in NURBS can be performed well using Boehm's algorithm [51], Oslo algorithm [52] or the blossoming principle stated by Goldman [53]. Knot insertion works well for curves when only few control points are added around the region. However, this does not work for tensor product B-spline surfaces because insertion of a single knot vector causes an addition of control points in the entire row or column and it is also not possible without changing the shape of the geometry. The reason for this limitation is because of its tensor product construction as shown in the equation 1.1. The only possibility left behind is representing the optical surface by an alternate form without any change in its shape.

5.2 T-SPLINES

The limitations imposed by NURBS can be solved by an alternate surface representation called T-Splines. T-Splines are an advancement of NURBS that allows the surface to create T-junctions. These junctions allow T-Splines to refine locally which means the control points can be inserted without propagating an entire row or column of control points [54]. The main advantage of the T-Splines is that their local refinement does not influence the shape of the curve or surface. T-Splines also offer another advantage. The unwanted control points which carry no information can be removed because the superfluous control points are really a disturbance to the designer. This section covers the concepts behind the T-Splines.

THEORY

T-Splines can be generalized as tensor product B-Splines which is point based instead of grid based [54]. The control grid is called T-mesh. If a T-mesh forms a rectangular grid, then it reverts back to NURBS. A T-spline can be defined as,

$$S(u,v) = \frac{\sum_{i=0}^{n} \boldsymbol{p}_i w_i N_i(u,v)}{\sum_{i=0}^{n} w_i N_i(u,v)} \tag{5.1}$$

where $\boldsymbol{p_i}$ are control points, w_i its weights and $N_i(u,v)$ are the basis functions which could be represented as,

$$N_i(u,v) = N_{ui}(u) \cdot N_{vi}(v) \tag{5.2}$$

The knot vectors associated with these basis functions $N_{ui}(u)$ and $N_{vi}(v)$ are given by,

$$\boldsymbol{u_i} = [u_{i0}, u_{i1}, u_{i2}, u_{i3}, \ldots, u_{in}], \quad \boldsymbol{v_i} = [v_{i0}, v_{i1}, v_{i2}, v_{i3}, \ldots, v_{in}] \tag{5.3}$$

T-MESH

In T-Splines, the knot interval information for the basis function is conveyed using a lattice called T-mesh or T-grid [50]. The figure 5.3 shows an example of a T-mesh in knot coordinate (u,v) associated with knot intervals d_i and e_i [54]. The line segments in u direction and v direction are defined as edges of the T-mesh. The set of edges connected by T-junctions forms a face denoted as $\boldsymbol{F}$ in the figure 5.3. The sum of the knot intervals on both the sides of the face $\boldsymbol{F}$ must be equal. For example the face $\boldsymbol{F}$ in the figure 5.3 must satisfy $e_7 + e_6 = e_8 + e_9$. The control grid point $\boldsymbol{P_1}$ lies at (u_2, v_2) and the control point $\boldsymbol{P_2}$ lies at $(u_3 + d_6, v_3)$ respectively.

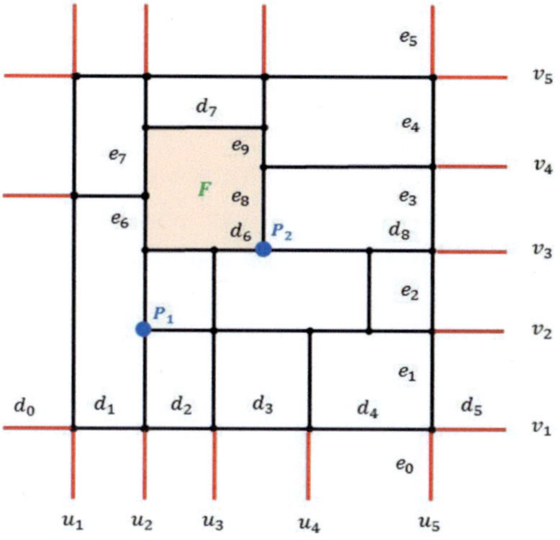

Figure 5.3: An example of a T-mesh showing the knot vector representation $\boldsymbol{u_i}$ and $\boldsymbol{v_i}$, and knot intervals d_i and e_i.

KNOT INSERTION AND BASIS FUNCTIONS REFINEMENT

Knot insertion is used to add detail to the needed region in the optical surface. This leads to the local refinement of the T-Splines without altering the shape of the freeform surface [48]. This is because the refinement is carried out for the parametric directions $\boldsymbol{u}$ and $\boldsymbol{v}$ independent of each other unlike in NURBS. The refinement of the basis functions $N_i\,(u,v)$ is done separately using two individual univariate basis functions $N_{ui}\,(u,v)$ and $N_{vi}\,(u,v)$ respectively by fulfilling the equation 5.2. The procedure about the refinement of basis functions when a new knot vector gets inserted is illustrated in the appendix B using a simple example. This property of local refinement without any increase in the number of control points as well as no change in the shape of the geometry makes T-Splines naturally a good choice to implement in the OFFD to attain local deformation.

5.3 APPLICATION OF T-SPLINES IN OFFD

This section presents the application of the T-Splines in the OFFD. To analyze its performance and compare against NURBS, the task of designing a street light lens introduced in the chapter 3 is used. The initial system, the lighting requirements and merit functions to evaluate the photometric performance remain the same.

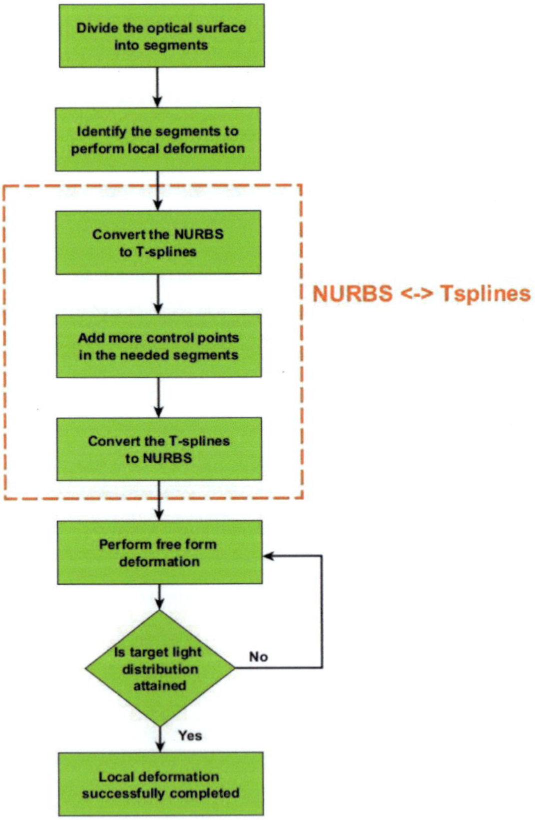

Figure 5.4: Workflow of the T-Splines implementation in OFFD.

5.3.1 WORKFLOW

The workflow that shows the implementation of the T-splines in the OFFD is summarized as a flowchart in the figure 5.4. To begin with, the initial surface must be divided into segments based on the designer's needs. The segments to perform local deformation have to be identified and the optical surface represented as NURBS must be converted to T-Splines. The control points are then increased in the segments where local deformation is aimed. Due to its topological representation, the area and the shape of the surface remain unchanged.

More local deformation is pronounced in the regions where the density of control points is higher. After more control points are added to the selected sections, the optical surface is converted back to NURBS. This is necessary because NURBS is the compatible format for the OFFD but the modification made is well preserved. The freeform deformation routine is carried out later using this modified surface until the target lighting requirements are attained.

5.3.2 VALIDATION

The whole optical surface is divided into many sections as shown in the figure 5.5. By taking the symmetry of the street into account, six segments are used in this analysis. For a straight forward comparison, the OFFD grid points $[1, 3, 13, 15]$ as represented in the figure 4.1 are chosen. The number of control points is doubled in each segment. In the end, six new optical surfaces are generated.

The only difference between these generated new T-spline surfaces and the initial one is the difference in the number of control points. The shape of the optical surface remains the same without any variation as expected. This study is mainly carried to find which segment of the lens requires more local deformation that ultimately could lead to the improvement in the optical performance.

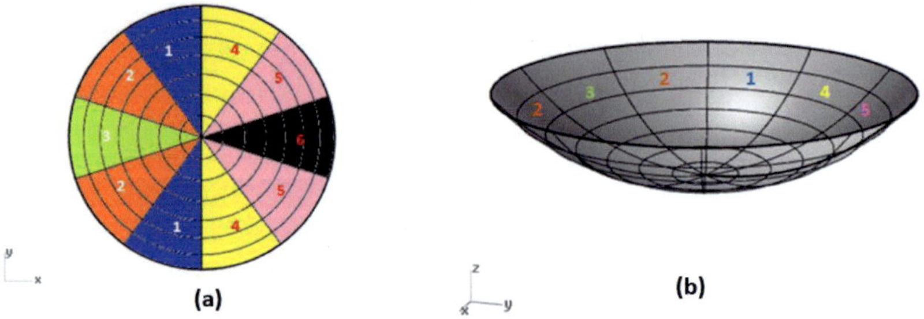

Figure 5.5: Initial optical surface segmented into six sections shown in (a) top view and (b) perspective view.

Section	ΔQ_{dev}	$\Delta\eta$
1	86.5	39.7
2	86.5	36.1
3	88.0	33.6
4	86.4	38.6
5	**89.1**	**39.8**
6	76.5	32.3

Table 5.1: Table showing the improvement in the light distribution ΔQ_{dev} and improvement in efficiency $\Delta\eta$ after applying deviation based and flux based merit functions individually for each sections from 1 to 6.

These six optical surfaces are taken as initial surfaces for the OFFD and are evaluated one after the other using the deviation based and flux based merit functions. The preliminary results as seen in the table

5.1 show that the OFFD could provide an improvement in the light distribution and optical efficiency when more control points are added to the segment 5. So this optical surface with more control points at segment 5 and less at the remaining sections is taken as an initial system for optimization and also for comparison against NURBS and the results are discussed in the coming section.

5.3.3 COMPARISON BETWEEN NURBS AND T-SPLINES

The two important photometric measures used in analyzing the street lighting lens are total luminous flux in the targeted region and the shape of the distribution. NURBS and T-Splines performed same in maximizing the total luminous flux in the needed region. But T-Splines outperformed in shaping the light distribution as required which is validated in the figure 5.6. The illuminance distribution using T-Splines is more uniform compared to NURBS.

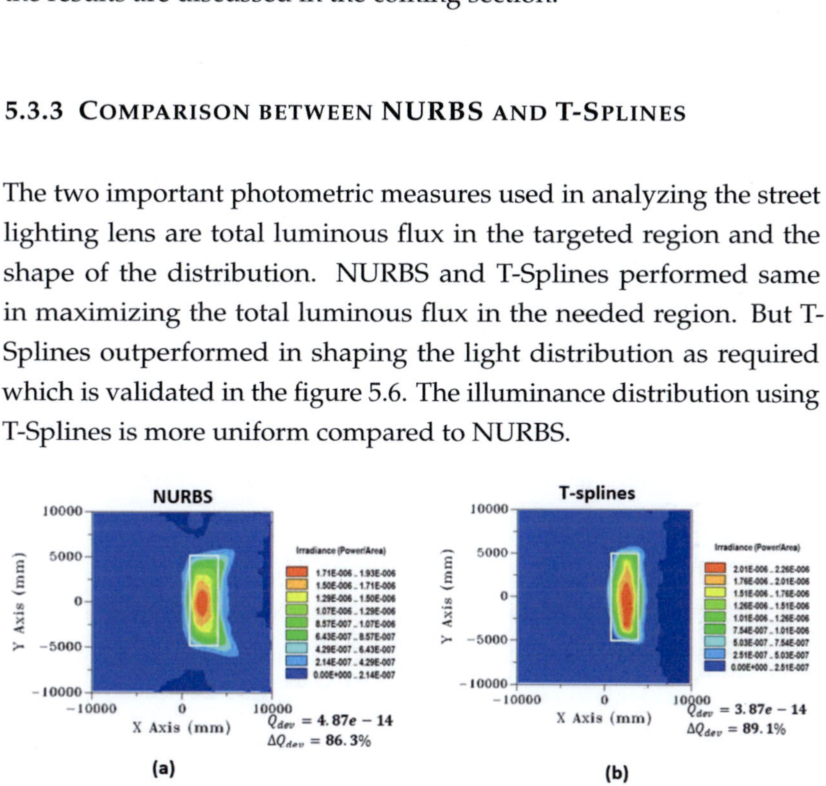

Figure 5.6: Illuminance distribution of the streetlight lens optimized with added control points on the fifth section using deviation based merit function (a) NURBS (b) T-Splines.

The deformed optical surfaces using NURBS and T-Splines are shown in the figure 5.7. The slight difference is seen in the T-Splines near the edges which got mapped as segment 5 in the figure 5.5. This is the segment where more control points have been added prior to deformation and performed better than the others.

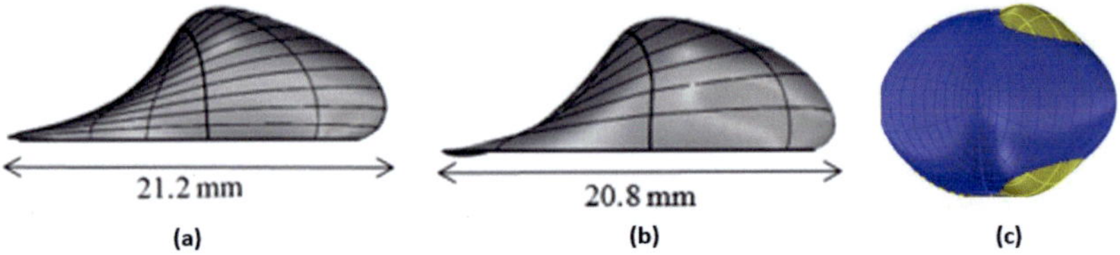

Figure 5.7: Deformed optical surface using (a) NURBS based OFFD, (b) T-Splines based OFFD and (c) false color representation showing the change in shape between the NURBS and the T-Splines.

If the same results need to be attained using NURBS, accurate control points in the grid must be selected because it is more sensitive to the designer's choice. Even after it is accurately chosen, the same results can be attained only at the expense of the optimization's run-time which is almost twice the time needed using T-Splines for this particular street lighting application.

The most significant advantage the T-Splines could offer is that when the optical designer knows the section of the surface to be deformed in advance, he could add more control points locally in the region and this modification reduces the sensitivity of the OFFD grid. This is because the optical designer has interacted closely with the surface by taking advantage of the surface's property and modified the surface even before the optimization begins.

Later when the optimization is carried out, the impact of the selected control grid points to be as inputs to the optimization has minimal role in the optical performance as compared with NURBS where prior modification of the surface is impossible.

5.4 CONCLUSION

This chapter highlighted the use of T-Splines by performing local deformation of optical surfaces. The T-Splines have been implemented and its optical performance was analyzed for the first time. This approach was made possible because of its seamless interoperability between NURBS and T-Splines. The control points could be precisely added in the required regions with ease using T-Splines. The results showed that the OFFD using T-Splines attained more uniform light distribution compared to NURBS. It clearly indicates that more illumination and architectural lighting problems could be easily addressed using T-Splines in the near future. Moreover, the sensitivity of the OFFD grid points got reduced when T-Splines based OFFD was used.

As T-Splines are more advanced form of surface representation, they are not yet matured. The CAD techniques and raytracers have not grown to an extent to import and work directly with the T-Splines file format. So at the time of this work, one needs to still rely on conversion of T-Splines to NURBS to perform raytracing. If raytracing for optical surfaces using T-Splines is available, then significant iterative refinement on the control regions can be made in the optimization system itself. Another limiting factor but not huge is that the control points have to be added more precisely in the needed regions. In future, if intelligent optimization systems could predict the segments of the optical surface where modification is necessary, then more control points can be added in the respective region automatically and optimization will be then carried out.

CHAPTER 6

MANUFACTURING FEASIBILITY ANALYSIS

An optical designer enjoys his freedom of creativity as any surface can be mathematically represented. But his freedom to realize them is limited by its ability to manufacture. Unfortunately manufacturing feasibility analysis is not yet implemented in the OFFD. By incorporating them into the OFFD, it provides two benefits. First, a manufacturable surface is created and the second one is the improvement in the timing efficiency of OFFD as unwanted surfaces are not further computed.

6.1 SIGNIFICANCE

Knowledge about tolerances is important for manufacturing optical surfaces [2]. In general, any surface must exhibit continuity of position, tangent and curvature. It should be free of extraneous bumps or wiggles or unwanted inflections [55]. Many surfaces in CAGD derive new surfaces from initial ones subject to certain restrictions. Examples include offset surfaces, where a distance constraint must be fulfilled and spherical blends, where a curvature constraint must be satisfied [56]. Therefore, a precise method to evaluate the quality of the surface is mandatory during the design phase of any optical system.

In OFFD, the optimization provides shifts to the selected control grid points along its three directions (x, y, z). But the direction and the extent it could traverse on the grid's landscape is not restricted. This leads to the creation of many surfaces that cannot be manufactured. Some of the results produced by the OFFD are shown in the figure 6.1. This challenges the optical designer as the final optimized result is not always guaranteed to be manufactured. When such results are obtained, he needs to inspect manually until he finds the next possible surface suitable to manufacture.

(a) (b) (c)

Figure 6.1: Examples of infeasible optical surfaces $(a) - (c)$ generated by the OFFD during the street lighting lens optimization.

In worst conditions, when irrelevant grid points are selected, OFFD generates wildly deformed surfaces at the very beginning of the optimization. Since it could not detect intelligently, the rest of the system follows by modifying this infeasible surface making it even worse. So at the end of the optimization after a long wait, the designer has a surface which fulfills the photometric requirements. But this can be neither used for further improvement (as a next iteration for the optimization) nor for manufacturing.

So this chapter aims to solve this problem by implementing a suitable surface analysis tool. It analyzes the size and curvature of the generated optical surface and restrict infeasible surfaces by placing suitable limits in the OFFD. This newly implemented method is validated using street lighting lens and automotive stop lamps introduced in the chapter 3. Finally, this chapter ends with a comparison analysis between the previous [4] and this current implementation.

6.2 BACKGROUND AND THEORY

The mathematical background required to manipulate the curvature and the offset of the NURBS surfaces are presented in this section.

CURVATURE ANALYSIS

When any surface is presented to the designer, he is likely to encounter terms like "too flat, unwanted round edges at the top, folded back" etc. by visual inspection. For a computer to deal with them, these shape terms must be translated into mathematical form [57]. Mathematician Euler was the first to propose the term curvature in 1760. According to him, the curvature of any curve or surface is equal to the magnitude of the second derivative of a curve or surface at a particular point [58, 59]. Later on, Gauss recognized the importance of curvature and made it popular. He claimed that it can be used to describe any thing in space mathematically independent of the coordinate system. The curvature is now popularly used in the automotive, aircraft and graphics industry as they need to describe and modify complicated geometries using CAGD systems. In Euclidean three-dimensional space, the curvature of any parametric NURBS based surfaces can be defined using Gaussian curvature [60] which helps to show the anomalies in the surface like bumps, dents, ripples, etc. The computation of the Gaussian curvature is explained in the coming section.

GAUSSIAN CURVATURE OF A SURFACE

For an object in three dimensional space, there is always a tangent plane to that surface at a specific point (u, v). The normal curvature sections can be then computed for all directions to this tangent plane. From these, two curves crossing this point (u, v) are selected, the one with a minimum curvature K_{min} and the other with a maximum curvature K_{max} as shown in the figure 6.2. The Gaussian curvature denoted by κ is the product of K_{min} and K_{max} [61].

$$\kappa = K_{min} \times K_{max} \tag{6.1}$$

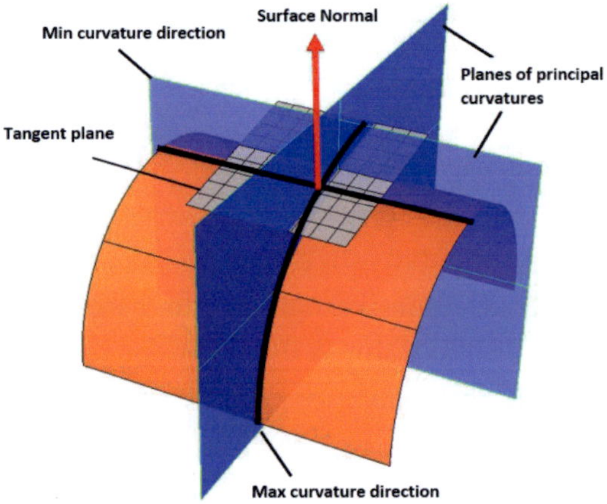

Figure 6.2: Schematic representation of the NURBS (orange) with a tangent plane (grey) being cut by the planes of principal curvatures (blue).

Inflection occurs at a point where the surface meets its tangent plane. If the surface exhibits positive curvature, the tangent plane touches the surface at a point (u, v). If it shows negative Gaussian curvature, then the tangent plane cuts the surface at a specific point (u, v). This is why a surface resembles like a saddle for negative Gaussian curvature, flat for zero Gaussian curvature and looks like a bowl when it exhibits positive Gaussian curvature as seen in the figure 6.3.

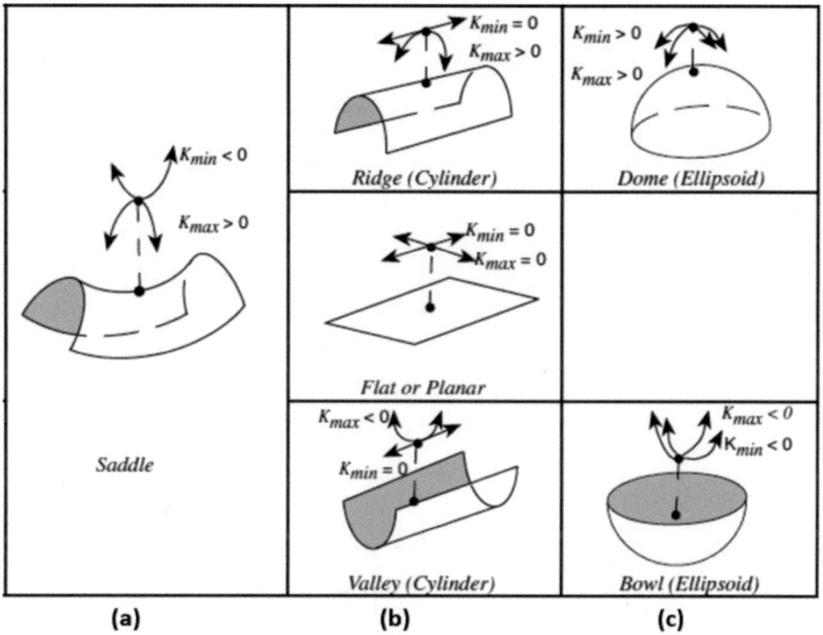

Figure 6.3: Examples of surfaces showing (a) negative Gaussian, (b) zero Gaussian and (c) positive Gaussian curvature [62].

GEOMETRIC INTERPRETATION OF GAUSSIAN CURVATURE

In this section, the computation of Gaussian curvature for NURBS and its interpretation [60][61][57][63][64] have been summarized.

Let K_{min} and K_{max} be the principal curvatures of a regular tensor product parametric B-Spline surface $S(u,v)$. The Gaussian curvature of this surface under investigation $S(u,v)$ is given by,

$$\kappa(u,v) = K_{min}(u,v) \times K_{max}(u,v) \tag{6.2}$$

To compute κ, the first and the second fundamental forms of the surface must be computed. The surface normal is given by $n(u,v) = \frac{S_u \times S_v}{\|S_u \times S_v\|}$, with $\|S_u \times S_v\| \neq 0$ since S is regular. The subscripts S_u and S_v indicate the partial derivatives with respect to the corresponding parameters u and v respectively. The matrix representation of the first fundamental form of the surface is given by,

$$I = \begin{bmatrix} E & F \\ F & G \end{bmatrix} = \begin{bmatrix} \langle S_u, S_u \rangle & \langle S_u, S_v \rangle \\ \langle S_u, S_v \rangle & \langle S_v, S_v \rangle \end{bmatrix} \tag{6.3}$$

The matrix representation of the second fundamental form is given by,

$$II = \begin{bmatrix} L & M \\ M & N \end{bmatrix} = \begin{bmatrix} \langle S_{uu}, n \rangle & \langle S_{uv}, n \rangle \\ \langle S_{uv}, n \rangle & \langle S_{vv}, n \rangle \end{bmatrix} \tag{6.4}$$

Let A, B, C be defined as follows

$$A = EG - F^2 \tag{6.5}$$

$$B = 2FM - GL - EN \tag{6.6}$$

$$C = LN - M^2 \tag{6.7}$$

Now the principal curvatures at a point on the surface is given by,

$$K_{min} = \frac{-B-\sqrt{B^2-4AC}}{2A} \quad ; \quad K_{max} = \frac{-B+\sqrt{B^2-4AC}}{2A} \quad K_{max} \geq K_{min} \tag{6.8}$$

Then the Gaussian curvature is a product of these two principal curvatures K_{min} and K_{max} .

$$\kappa = K_{min} \times K_{max} = \left(\frac{-B - \sqrt{B^2 - 4AC}}{2A} \right) \left(\frac{-B + \sqrt{B^2 - 4AC}}{2A} \right) \tag{6.9}$$

By substituting the suitable partial derivatives for A, B and C and solving the equation, it gets simplified to

$$\kappa = \frac{LN - M^2}{EG - F^2} \tag{6.10}$$

In NURBS, $\kappa(u, v)$ has to be evaluated as a function value for each section of the NURBS patch whose interval is limited by the knot vectors. For a NURBS with n knot vectors along u direction and m knot vectors in v direction, the Gaussian Curvature has to be evaluated for a total of $n \times m$ points. By doing so, the region of the NURBS where it has undergone deformation is known. More importantly, the magnitude of the computed result tells the quantitative amount of curvature being pronounced on the surface. If the range of these curvature values is already known, they can be specified as tolerance limits to the OFFD. The information regarding the range of curvature values can be obtained from optical manufacturers. The curvature varies significantly depending on the manufacturing type and the materials used during the process.

OFFSET ANALYSIS

Offset analysis is very important when the overall geometry must be restricted to defined dimensions. This could be either diameter or thickness or even overall size of an optical surface. This will be also helpful to define minimum or maximum allowable source-optics distance. These parameters can be evaluated directly using any CAD tools. But in an automated optimization, these requirements have to be specified indirectly as tolerance limits to the OFFD. By specifying these tolerances, the arbitrary change in the shape of the freeform surface can be restricted. For example, if thickness(t) and diameter (d) of the optical surface shown in the figure 6.4 has to be confined to particular dimensions, then the control points in the z direction must be limited to any defined thickness and control points along y for the diameter. This is a simple case taken as an example to show the importance of offset analysis. This analysis is much more significant during the optimization of the compact optics as it can be used to control the optical surface.

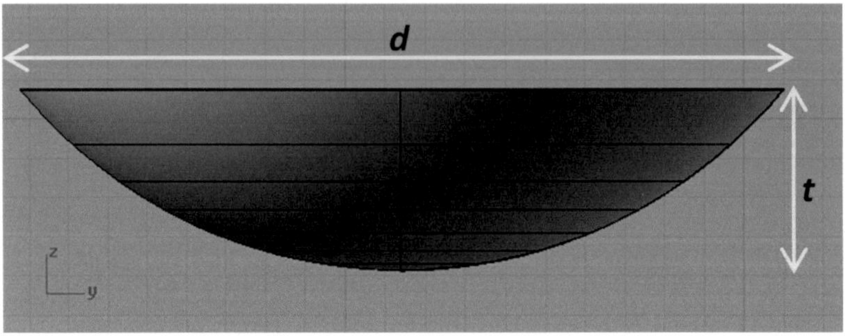

Figure 6.4: Spherical surface with diameter *d* and thickness *t*.

6.3 IMPLEMENTATION

As the theoretical background and its significance are understood, the next step is to implement and integrate them into the OFFD and evaluate its performance. The curvature analysis tool is created based on the formulas from the section 6.2 and they are encapsulated as a toolbox. The tolerance limits regarding the optics size are given as inputs to the OFFD. The main difference between the previous OFFD and this work is the addition of the manufacturing feasibility analysis block shown as yellow in the flowchart shown in the figure 6.5.

The inputs required for the OFFD which include the control grid points, initial surface, photometric requirements and tolerance limits for the surface geometry must be specified. The system begins by deforming the OFFD grid based on the shifts it receives from the optimization. The change in the shape of the grid creates a new optical surface. Then this surface is analyzed for manufacturing feasibility.

To analyze the surface, the surface is divided based on its knot interval with n knot vectors along u direction and m knot vectors in v direction. Then the curvature and its size are analyzed for the entire domain and a $n \times m$ analysis matrix is created and this matrix is compared against the threshold values. If it is less or equal, then the surface is allowed for raytracing and further photometric evaluation. If it does not satisfy, then this surface is discarded and the algorithm re-iterates back to the optimization block to search new shifts for the grid control points. This trigger to search new shifts can be done by many ways.

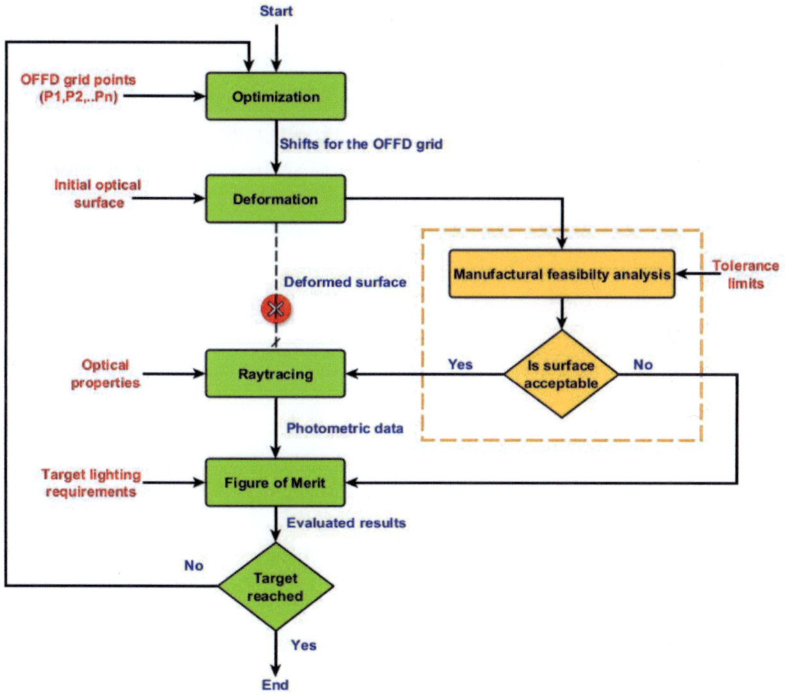

Figure 6.5: Workflow showing the OFFD with manufacturing feasibility analysis implemented into the system.

6.3.1 WORK FLOW

The first approach is straightforward and this can be done by adding constraints to the optimization algorithm. In any optimization algorithm, constraints can be imposed to the optimization variables by specifying appropriate bounds. This works well for direct optimization problems. As OFFD is an indirect optimization technique where the cubical grid is being modified by the optimization algorithm during the entire process, only the movement of the cubical grid points can be restricted. As the surface parameters which include size and curvature of the optical surface has no relationship with these cubic grid points, setting constraints to the cubical grid points is not possible. Therefore, constrained optimization cannot help OFFD to include manufacturing feasibility analysis into the system.

The second method to establish this interaction was implemented using a multi objective merit function. This merit function encompasses the contribution from the photometric measure (Q_{pho}) and the geometric measure (Q_{geo}) of the deformed optical surface and can be mathematically represented as in equation 6.11

$$Q = Q_{pho} + Q_{geo} \tag{6.11}$$

where Q is the summed value of both the photometric and the geometric measure of the optical system and ranges from 0 to 1. But unfortunately, this merit function did not improve the surface contrary to all expectations. This is because the evaluated Q is a result of two different requirements. An improvement in Q occurred either when there was an improvement in the photometric measure or improvement in the geometric measure. But the optimization did not know the roots of its total contribution (Q) whether it was due to Q_{pho} or

Q_{geo}. As this information cannot be communicated using a single numerical value, this misled the optimization process and yielded either unoptimized solutions or failed to converge in search of a non existing solution. Moreover, describing the geometric measure which consisted of various surface parameters like curvature, thickness and diameter in to a single accurate value is challenging. So describing the illumination problem which encompasses both the photometric and the geometric requirements using a single value is not feasible.

The last and the only method which showed an improvement in the performance is the modified form of the previous merit function described in the equation 6.11. It follows a non smooth penalty based binary approach in which the optimization is allowed to continue if the curvature and size are well within its limits.

If it exceeds, the value of the merit function governing them is increased to the maximum limit. This is done using the penalty parameters μ_{geo} and μ_{pho} which govern the geometric and the photometric requirements respectively. μ_{geo} remains zero and μ_{pho} is assigned to one if the surface is suitable to manufacture and Q is therefore as a result of its evaluated photometric measure (Q_{pho}).

When the system encounters any infeasible surface, the penalty parameter μ_{geo} is increased to one and as raytracing and photometric evaluation are not performed for such surfaces, the parameter μ_{pho} is assigned zero. By doing so, the Q reaches to one which is its maximum limit. The penalty based merit function is mathematically represented in the equation 6.12.

$$Q = \mu_{pho} Q_{pho} + \mu_{geo}; \quad \begin{matrix} \mu_{geo}=0 \; and \; \mu_{pho}=1; \forall \; feasible \; surfaces \\ \mu_{geo}=1 \; and \; \mu_{pho}=0; \forall \; infeasible \; surfaces \end{matrix} \tag{6.12}$$

This adds more physical meaning to its evaluation measure which includes the geometric and the photometric measure of the optical surface in an intuitive way.

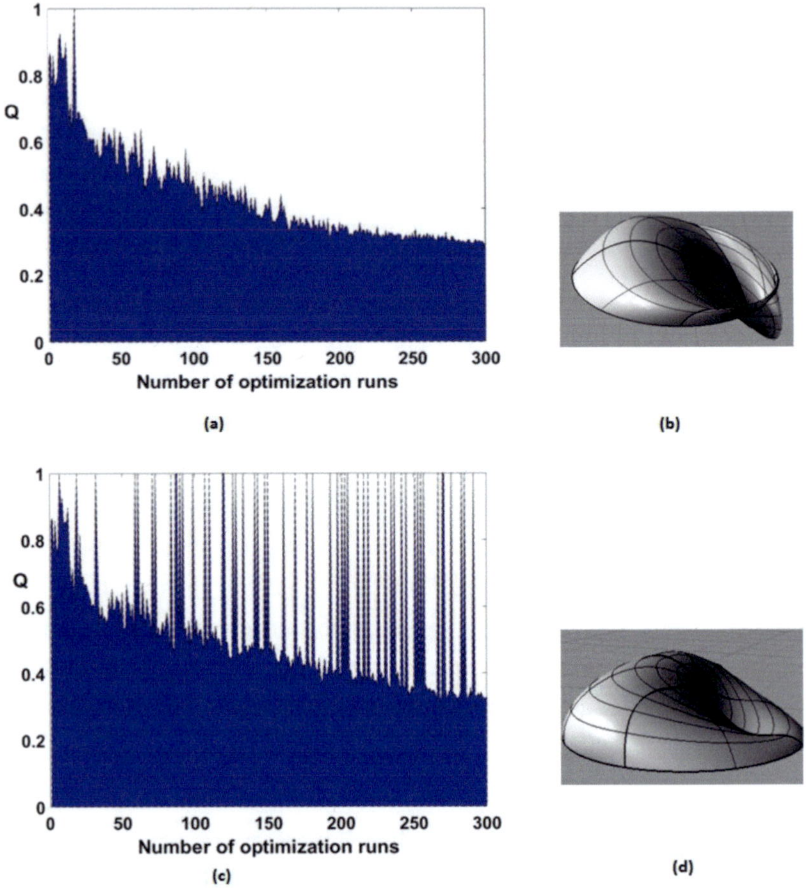

Figure 6.6: (a) Q Landscape of the evaluated merit function based only on the photometric evaluation used in previous work [4], (b) deformed infeasible street lighting lens obtained as a result, (c) Q landscape when penalty based merit function based on equation 6.12 is used and (d) deformed surface suitable to manufacture as well as satisfying the photometric requirements obtained as a result.

As the evaluated merit function is altered to a high value, this will be indicated to the optimization algorithm as worst performance and it searches differently to direct the optimization in the right direction. This is the same approach, an optical designer follows when he manually encounters such situation and this approach is transferred to the OFFD system. The OFFD thus interactively changes the surface and the computation ends when the targeted photometric characteristics are attained for a surface that is suitable to manufacture.

The figure 6.6(a) shows the landscape of the evaluated Q when the merit function is based only on the photometric performance as used in the previous work [4] and the infeasible surface generated as result of it is seen is the figure 6.6(b). Whereas, the figure 6.6(c) shows the landscape of Q when penalty based merit function as described in the equation 6.12 is used. The spotted peaks in the landscape are the instances when the infeasible surfaces are detected by the system and these surfaces are not computed further. The Q progression is continued until a surface satisfying both the photometric and geometric measures as shown in the figure 6.6(d) is obtained.

6.3.2 VALIDATION

In order to validate this analysis, the street lighting lens and the automotive stop lamps discussed in the chapter 3.1 are used.

STREET LIGHTING LENS

In the street lighting system, there are 12-knot vectors along u direction and 14-knot vectors along v direction and so the total number of critical points ($n_{critical}$) to analyze are $12 \times 14 = 168$ points. The maximum of the surface curvature denoted as $Curv_{max}$ is limited to

0.075 and minimum $(Curv_{min})$ should be above -0.050 respectively. The curvature limits are selected based on the experience but this is usually obtained from optical manufacturers.

In OFFD, the diameter (d) can be interpreted as the change of control points in the y direction. For thickness, it is the change in the difference between the location where the deformed surface ends z_{end} and where the deformed optical surface starts z_{start}. So these parameters can be restricted by placing suitable limits along y and z directions respectively. When calculated from the origin, z_{end} comes around $9.6mm$. So the threshold limits are set in such a way that the diameter and thickness of the optical surface do not enlarge further. If the deformation is large, then this deformed surface will not fit into the other surface of the street lighting lens as shown in the figure 6.7.

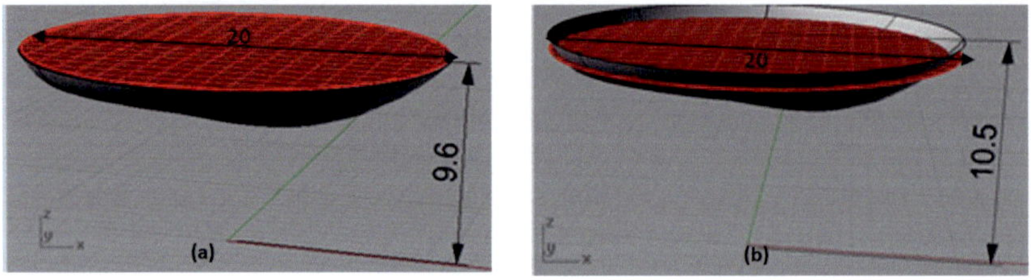

Figure 6.7: Deformed optical surface shown in grey (a) within the tolerated thickness and (b) beyond the limit making it infeasible to attach to the plain end surface (shown in red) of the street lighting lens system with dimensions in millimeters.

For a surface to be manufactured, the evaluation results should fall inside the defined threshold else the surface is discarded. The computational characteristics used as limits in evaluating the deformed surface for the street lighting lens are summarized in the table 6.1.

The results obtained during the optimization of the street lighting lens are shown in the figure 6.8 along with the evaluated numerical results at the bottom of each deformed surface. The first surface shown in the figure 6.8(a) is the initial spherical surface before optimization and exhibits positive curvature as expected. As the surface gets slowly deformed, it begins to exhibit more positive as well negative curvatures based on the deformation of the OFFD grid.

The surfaces from 6.8(a) – (c) are well within the limit and these surfaces could be manufactured which is evident when one examines these surfaces visually.

$n_{critical}$	$Curv_{min}$	$Curv_{max}$	$d\,(mm)$	$z_{end}\,(mm)$
168	-0.050	0.075	20	9.6

Table 6.1: Computational tolerances like curvature and size for the street lighting lens.

The optical surface in the figure 6.8(d) exhibits more positive Gaussian curvature leading to a bump in the surface and the evaluated result ($Curv_{max} = 0.079$) is above the threshold. The figure 6.8(e) exhibits negative Gaussian curvature ($Curv_{min} = -0.081$) leading to a back fold in the surface and diameter (d) exceeds its limit as well.

The optical surface in the figure 6.8(f) shows high positive Gaussian curvature ($Curv_{max} = 0.086$) and the thickness calculated indirectly as z_{end} is above the threshold ($z_{end} = 10mm$) where the maximum allowed is only 9.6mm.

As the deformation of the grid becomes stronger and stronger, its impact on the optical surface gets higher which is evident from figures 6.8(g) – (i) where the positive or negative curvature increases tremendously and a substantial increase in the thickness too.

The surfaces 6.8(d) – (i) are discarded during the optimization as the evaluated parameters lie beyond the threshold. During this situation, the OFFD re-iterates back to get a feasible surface by finding alternate shifts from the optimization algorithm.

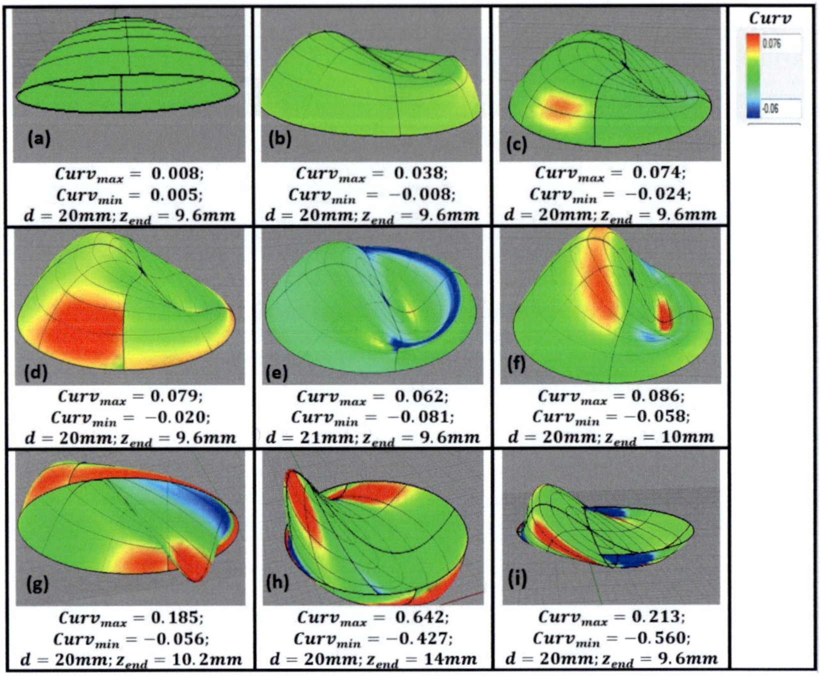

Figure 6.8: Examples of deformed surfaces for the street lighting system obtained as a result of the manufacturing tolerances implementation in the OFFD. The optical surfaces at the first row (a) to (c) are feasible to manufacture as the curvature and the size are within the tolerance limits. The surfaces from (d) to (i) are infeasible surfaces exhibiting more positive (red) and negative (blue) curvatures and the thickness exceeding its limit.

STOP LAMPS FOR AUTOMOTIVE LIGHTING

The second optical system used to analyze the performance of this new implementation is the automotive stop lamp. Unlike street lighting lens, the hybrid TIR optics consists of more surfaces as shown in the figure 3.7 that are subjected to deformation during each OFFD iteration. But each surface has different curvature and size that must be evaluated during the optimization. So every surface is analyzed individually and carefully whether the curvature or its size exceeds the threshold. If it is within the limit, the optimization continues further else it re-iterates back.

Surface	$n_{critical}$	$Curv_{min}$	$Curv_{max}$	$d\,(mm)$	$z_{end}\,(mm)$
Refracting	324	-0.050	0.090	-	11
TIR	180	-0.001	0.010	28	11
Side entry	48	0	0	-	-

Table 6.2: Computational tolerances for each surface of the TIR hybrid optics of the automotive stop lamp.

Based on the number of knot vectors across the surface, the number of critical points $n_{critical}$ differ for each surface. The side entry surface as seen in the figure 3.7 is flat and uniform and hence less number of knot vectors are enough to represent the surface and therefore fewer evaluation points. But either the TIR or the refracting surface is larger and require more knot vectors to represent them accurately. Therefore more critical points are needed to evaluate the behavior of the surface. As multi surface deformation was carried out for the TIR optics, the diameter (d) of the refracting surface and the side entry surface does not vary significantly. Therefore, they are not considered for analysis but if the application demands restricting the diameter of the refracting surface to a certain threshold, then this must be carefully considered. However, in order to avoid the overall increase in the size of the TIR optics, the diameter of the TIR surface is limited to 28mm. The thickness is also kept under check by limiting z_{end} to 11 mm for both TIR and refracting surfaces.

The computational characteristics that are used as limits while evaluating the deformed surfaces are summarized in the table 6.2. Some surfaces that are analyzed and detected using this tool are shown in figures 6.9, 6.10 and 6.11. The curvature and its size are evaluated separately for TIR, side entry and refracting surfaces.

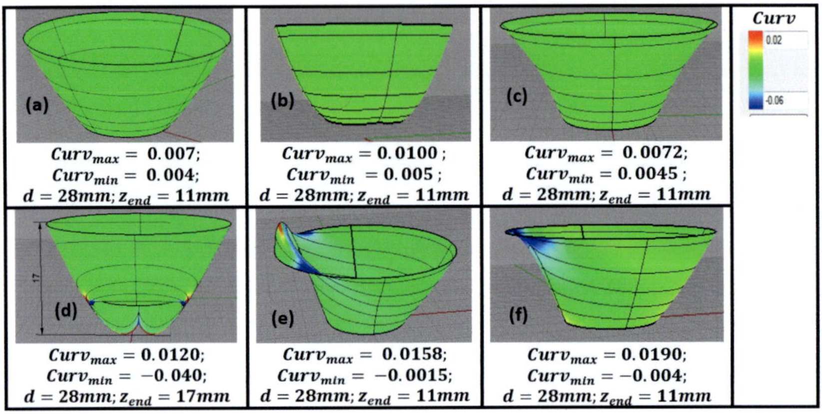

Figure 6.9: Examples of deformed TIR surfaces of the stop lamp obtained as a result of the manufacturing tolerances implementation in the OFFD. The optical surfaces at the first row (a) to (c) are feasible to manufacture as the curvature and the size are within the tolerance limits.. The surfaces from (d) to (f) are infeasible surfaces exhibiting more positive (red) and negative (blue) curvatures and the thickness exceeding its limit.

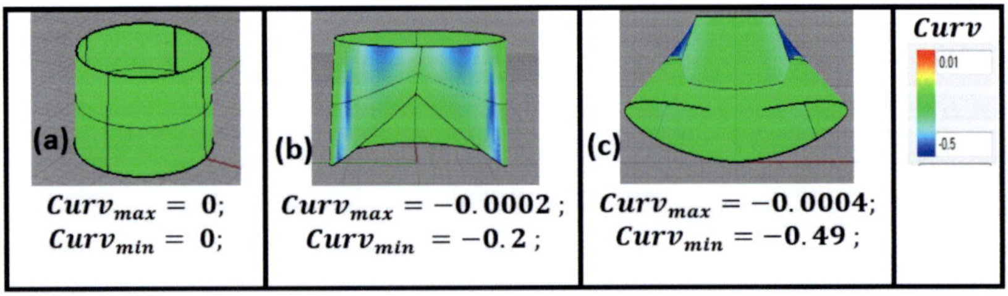

Figure 6.10: Examples of deformed side entry surfaces of the stop lamp obtained as a result of the manufacturing tolerances implementation in the OFFD. The optical surface (a) is feasible to manufacture as the curvature and the size are within the tolerance limits. The surfaces (b) and (c) are infeasible surfaces as they are exhibiting more negative (blue) curvature.

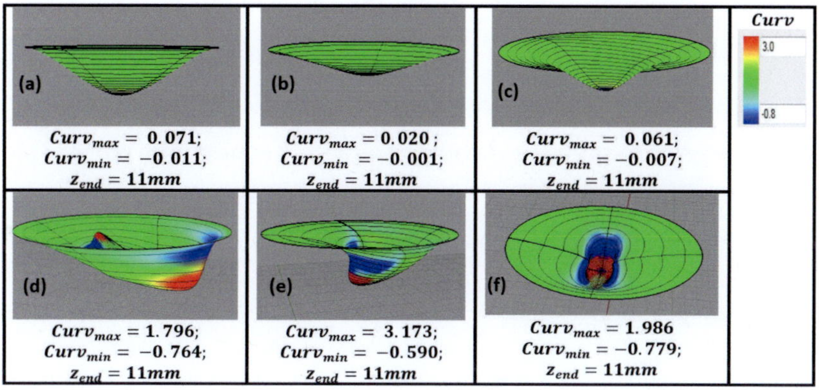

Figure 6.11: Examples of deformed refractive surfaces of the stop obtained as a result of the manufacturing tolerances implementation in the OFFD. The optical surfaces at the first row (a) to (c) are feasible to manufacture as the curvature and the size are within the tolerance limits. The surfaces from (d) to (f) are infeasible surfaces exhibiting more positive (red) and negative (blue) curvatures and the thickness exceeding its limit.

In figure 6.9, the TIR surface undergoes deformation and the optical surfaces from (a) to (c) are feasible to manufacture. The surfaces from (d) to (f) are infeasible surfaces as it gets either enlarged beyond the tolerance limit or folded back. These surfaces must be discarded even before the raytrace gets executed.

The deformation of the side entry surface is shown in the figure 6.10. Though the surface is flat at the beginning of the optimization as seen in 6.10(a), the deformation process modifies it as seen in 6.10(b) and 6.10(c) making it infeasible to manufacture.

Finally, some examples of refracting surface is shown in the figure 6.11. It is seen that the strong deformation is pronounced in the center of the refracting surfaces. As expected, the tool identified these infeasible surfaces 6.11$(d) - (f)$ when the curvature or size reaches its threshold. The tool also helped to identify the surfaces that are folded backwards making it infeasible to manufacture.

6.3.3 COMPARISON ANALYSIS

As a comparison analysis, this new implementation of analyzing the surfaces prior to raytrace is compared with the previous work [4] by taking street lighting lens as an example for its discussion.

MANUAL INTERVENTION

The manufacturing feasibility analysis has been encapsulated as a tool box in the OFFD and the threshold for offset and curvature are given as inputs to the OFFD system. This makes the OFFD generate results that satisfy not only photometrically but also guaranteed to be manufactured. This helps the optical designer to a greater extent as no manual intervention is required anymore either during or after the optimization to identify infeasible surfaces.

OPTICAL PERFORMANCE AND SPEED

The second advantage it could offer is an increase in speed. For the street lighting lens, the previous OFFD [4] required 1500 iterations by selecting four different combinations of grid points leading to an efficiency improvement ($\Delta\eta$) of 30.2%. As used in the previous chapters, the optimization was performed using deviation based merit function for 300 iterations followed by a flux based function for 300 iterations. But due to this new implementation, 116 surfaces are detected as infeasible ones and are not further computed. Therefore, it required only $600 - 116 = 484$ iterations to obtain an optimized surface. When the number of iterations are reduced, the timing efficiency also improves.

In a standard 4 core 3.30GHz machine, the previous OFFD required 209 minutes to finish the entire process with manual intervention to detect infeasible surfaces. Whereas the current implementation requires just 52 minutes to complete the entire design task using the same machine. Therefore to attain the same optical efficiency, OFFD with manufacturing feasibility analysis tool requires just one-fourth of the time required by the previous OFFD. The results are summarized in the table 6.3. The optical surfaces obtained from previous (as turquoise) and current (red) by overlapping with each other are shown in the figure 6.12 and both of them look alike thereby verifying that this current technique could find the best optical surface autonomously at a faster rate.

Parameters	Previous work [4]	Current results
Autonomous	No	Yes
No.of iterations	1500	484
Grid combinations	$(1,3,4,6,14)$, (2), (5), $(2,3,11,12)$	$(1,2,3,4,13)$
Efficiency improvement ($\Delta\eta$)	30.2%	30.2%
Time required (minutes)	209	52

Table 6.3: Comparison chart showing the results of the previous and the current OFFD after implementing manufacturing feasibility analysis in the system by taking street lighting lens as an example.

The reason for the longer duration of the previous work stems from the step size δ used in the OFFD. This δ determines the amount of shift of the optimization variables. If δ is small, the OFFD grid points are pushed or pulled to a smaller extent and hence a small change in the deformation of the optical surface. If δ is large, the OFFD grid points are displaced greatly leading to a significant change in the shape of the deformed surface. As no manufacturing feasibility analysis is present in the previous work, δ needs to be kept small, else the OFFD yields infeasible results at a faster rate. To manage this situation, δ is always kept small to have a controlled deformation but this costed time and it reached slowly to attain the desired optical performance. By having manufacturing feasibility analysis in the OFFD, the optical designer is free to choose his own δ. Even if he chooses a larger δ, infeasible optical surfaces are detected and the optimization is redirected accordingly.

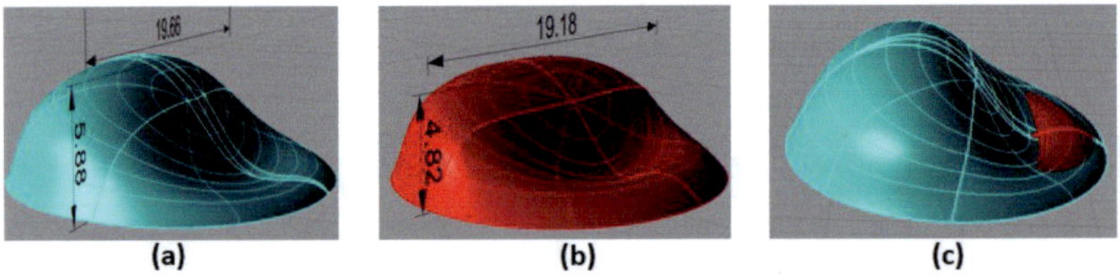

Figure 6.12: Deformed surface by (a) previous OFFD, (b) current with manufacturing analysis and (c) false color representation showing the change in shape between them with all dimension in millimeters.

CUSTOMER SPECIFIC NEEDS

As the manufacturing tolerances are given as inputs to the OFFD, it can be easily redefined based on each customer specific needs. Like for example, the curvature implemented in the OFFD will be useful in selecting the cutting size of the tool to avoid gouging (tool getting deeper into the surface than intended) in CAM systems [65]. And this can also be worked on the other way round too. When the manufacturing tool radius is known, the optical surface can be restricted to the minimum and maximum curvature by specifying as threshold limits to the OFFD. The offset analysis helps to impart the needed dimensions (diameter or thickness, source-optics distance or overall size) of the customer and obtain results accordingly.

6.4 CONCLUSION

This chapter showed the implementation of manufacturing feasibility analysis in the OFFD making it more advanced. By doing so, the optimization is being carried by considering the surface feasibility in one hand and photometric requirements on the other hand. Therefore the need of the manual intervention is eliminated. The needs of a customer can be imparted directly during the design phase thereby reducing the post manufacturing effects. The speed has been improved because larger deformation step size is used. This is more helpful when the performance of the initial optical surface is far from the target. During this situation, the optimization algorithm searches widely to reach the target leading to infeasible surfaces. But when surface feasibility analysis is present, its serves as a watch dog in discarding wildly deformed surfaces by re-iterating again till it satisfies target lighting as well the surface feasibility requirements. Most importantly, the time intense component in OFFD is the raytracing section. As infeasible surfaces are discarded by this current OFFD, the raytracing step is bypassed thereby attaining faster results.

The selection of control grid points also has an impact in improving the optical performance for the current OFFD. This issue will be addressed and a possible solution is discussed in the next chapter 7.

CHAPTER 7

INTELLIGENT DEFORMATION

In OFFD, the selection of appropriate grid control points for the optimization is challenging as grid points share no direct relationship with the optical performance. When inappropriate control points are selected, the computational complexity increases. This issue is addressed in this chapter by providing a suitable method to select the best grid points. By the end of this chapter, one could understand how far the OFFD has grown in terms of its intelligence.

7.1 SIGNIFICANCE

The important understanding that is missing so far is the relationship between the OFFD grid and the target lighting requirements. Since there is no one to one mapping between them, the selection of OFFD grid points was done using trial and error approach or the grid points were selected based on the prior experience. Moreover, the grid points are sensitive to target lighting requirements, initial design of the optics, and its type (refractive or reflective). So this selection was always dependent on the expertise of the optical designer. If he fails to select the optimal grid points, then complexity increases. This new approach overcomes these obstacles and selects the OFFD grid points on its own with more efficiency in a short amount of time. Intelligence must be imparted to OFFD by transferring the experience of an optical designer into machine understandable form using a set of decisions.

Every single operation and every decision by an optical designer are converted as tools in the OFFD so that it could find the optimal result on its own.

7.2 IMPLEMENTATION OF INTELLIGENT OFFD

The main objective is to create a tool which could find a relationship between its grid and the specified target requirements. When such relationship is established, the OFFD could find the best possible grid points. The optimization is guaranteed to attain the desired photometric requirements using these grid points in a short time. The method to find the best possible control points is explained in the following section.

7.2.1 CHALLENGES IN THE OFFD GRID

The 3D- OFFD grid as seen in the figure 7.1 can be represented using n control points marked as circles. A simple cubic grid can be constructed minimally using 27 grid control points and they can be selected as optimization variables in any arbitrary combination. So when all these 27 grid points are considered, the combinations of optimization variables calculated using nC_r where n is the total number of grid points and r is the number of grid points taken without repetition $r = (1, 2, 3, \ldots, 27)$ comes to the total of $27C_1 + 27C_2 + \cdots + 27C_{27} = 134,217,727$ combinations. To evaluate a single combination with 300 iterations, for a rayfile containing 1million rays in a standard 4 core 3.30GHz machine, it takes about 30 minutes. And when all the 134,217,727 combinations are evaluated, it takes 7600 years to get the best result if one follows a naive approach. In computational terms, these type of problems are called NP (nondeterministic polynomial time) hard and the correct solution can be attained only after infinite time. But a best possible solution can be found faster by providing intelligence or prior information to the system.

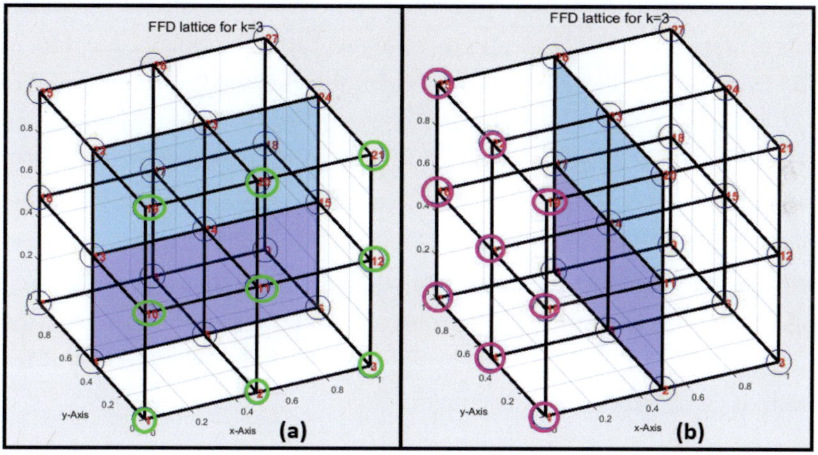

Figure 7.1: 3D- OFFD cubical grid showing (a) $y-$ axis symmetry with the grid points marked in green circles and (b) $x-$ axis symmetry with the grid points marked as magenta circles [4]. The grid symmetricity ensures that one half of the grid points are sufficient for the analysis and the other half can be mirrored.

7.2.2 SYMMETRY ANALYSIS

If there is any rotational symmetry in the target light distribution, then it gets simplified into a 2D design problem which can be solved using a two dimensional OFFD grid with only nine grid points. But most of the lighting distributions are complex and often do not fall in this category. So the use of 3D OFFD grid is inevitable. But in applications like street lighting lens, the street is symmetry along the $y-$ axis as seen in the figure 3.5 and in the stop lamp design, the intensity distribution is symmetry along both horizontal and vertical angles as seen in the figure 3.8. This symmetric property can be transferred into the OFFD grid. By doing so, the total number of combinations required for analysis gets reduced.

As seen in the figure 7.1, when the light distribution is symmetric along $y-$ axis, it is enough to analyze the grid points located at the front of the plane which are highlighted using green circles like $1, 2, 3, 10, 11 \ldots, etc$. The control points at the back are the mirrored version of the front and therefore not required to analyze them. Similarly, when it is symmetric along the $x-$ axis, the grid points at the left of the plane highlighted using magenta circles like $1, 2, 4, 10, 13, \ldots, etc$ are subjected to analysis as the control points at the right are the mirrored ones. So when the target distribution has either one symmetry, the total number of grid points to analyze gets reduced to half and for both, it gets reduced to one-fourth.

7.2.3 SHIFTING THE GRID AND SCANNING THE TARGET

In order to reduce the computational complexity, a suitable preprocessing technique has to be used. In OFFD, shift and scan method is used to eliminate the grid points that do not have significance in improving the optical performance. During this step, the relationship between the OFFD grid and the target distribution is established. This is done by shifting each grid point and evaluating its change in the light distribution. Usually, in any optimization, the obtained distribution is evaluated as a whole using a single numerical value Q. But this is inefficient to access its performance because the location where the modification is being done during the deformation remains unknown.

When an expert optical designer modifies a surface, he exactly knows where his modification could bring a change in the light distribution. This is the much-needed intelligence that has to be transferred to the OFFD in the first step.

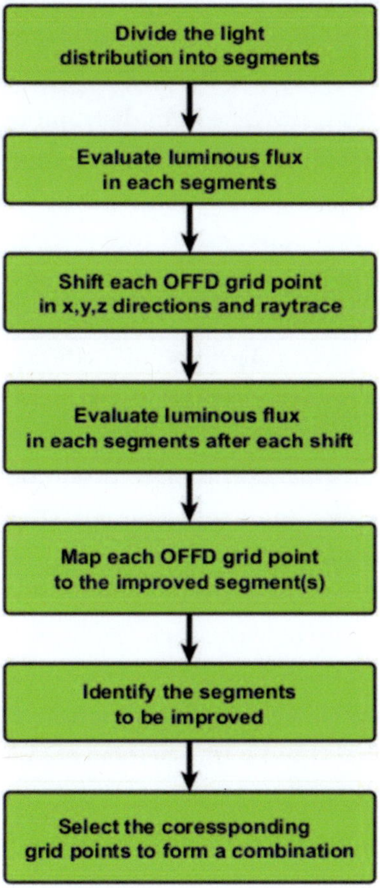

Figure 7.2: Flowchart of the preprocessing step implemented for the intelligent OFFD.

This can be achieved by dividing the light distribution into segments and evaluating the performance of each individual segment when the grid undergoes any change in shape. By doing so, the OFFD tool gains accurate control on the surface modification. When there is any shift in the OFFD grid point, its impact on the light distribution is now known spatially and quantitatively.

In OFFD, an initial surface and the target distribution are present already in the form of inputs. The light distribution generated by this initial surface will be analyzed and a better solution (set of OFFD grid points) to reach the target must be figured out. The proposed method as shown in the figure 7.2 works as follows: The light distribution created by the initial surface is divided into any arbitrary number of segments and the integrated luminous flux at each segment is calculated. Then each single OFFD grid point is shifted along the x, y and z directions respectively. This shift will be transferred to the underlying optical surface and this in turn will be reflected in the light distribution during the raytrace. During this step, it will be noticed that some segments perform better, some segments get worse and no change for others depending on the OFFD grid point. This performance measure at each segment in response to a particular OFFD grid point is stored. Thereby, the relationship between each OFFD grid point and the corresponding segment in the light distribution is established. From these results, the segments that need improvement and its corresponding OFFD grid points that could improve them are identified. These grid points form a combination and can be given as optimization variables to the OFFD later. As the OFFD has 27 grid points and when it is shifted along x, y and z directions, a total of $27 \times 3 = 81$ raytraces are enough to find the best possible grid points. In the case of symmetry, the total number of grid points and therefore the number of raytraces reduce to half. The total number of segments and its size are completely at the discretion of the optical designer and there is no specific restriction on it.

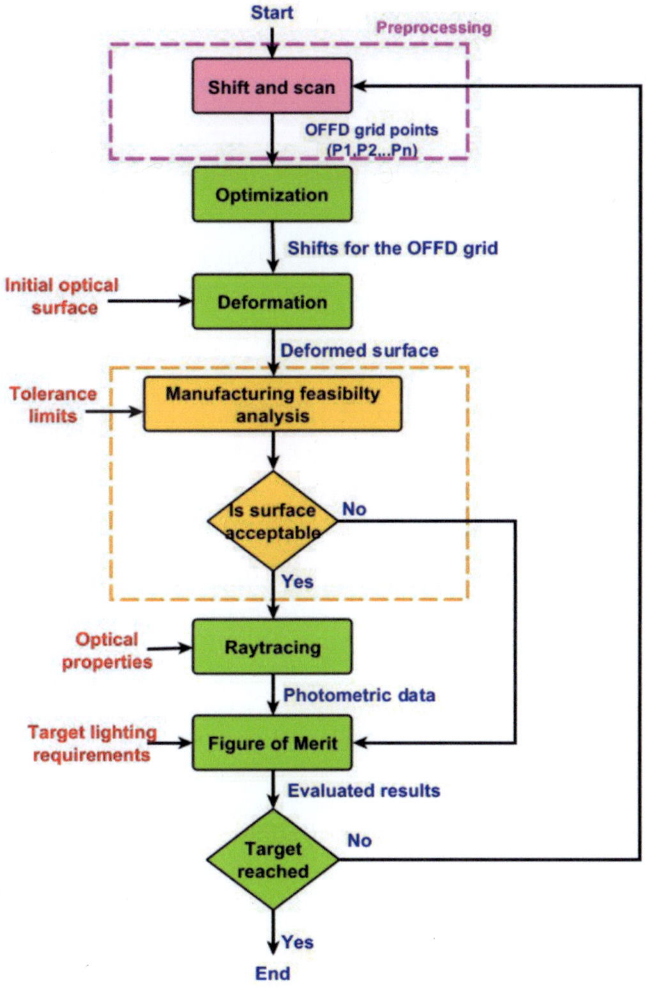

Figure 7.3: Flowchart showing the intelligent OFFD system with a newly integrated preprocessing step.

Segments can be divided small or large based on the lighting requirements and the optical performance of the initial system. If the objective is to attain homogeneous light distribution, then uniform segmentation is meaningful. For certain tasks like automotive applications, its angular lighting requirements differ throughout the region. For such requirements, non-uniform segmentation performs better. But care must be taken in selecting the resolution of the detector when segments are non-uniformly divided. In such case, the resolution of the detector must be adjusted suitably for each segment to avoid statistical noise. From now on, the initial guess on the grid points needed in the previous work is eliminated. The optical designer is not required to define the grid points as inputs anymore. The method to find optimal grid points has been integrated as a part of the OFFD by establishing a suitable grid-target relationship. This serves as a preprocessing process to the OFFD in selecting the appropriate grid points. The overall functionality is described using a flowchart in the figure 7.3 with the newly added module shown in pink. This new OFFD system is validated using street lighting lens and automotive stop lamps in the later sections of this chapter.

7.3 ARCHITECTURE OF INTELLIGENT OFFD

As a part of this dissertation, the architecture of the OFFD has been redefined significantly to improve the optical performance at a faster rate. The tool must be more flexible, user-friendly and extensible to add more functionalities in the future.

The Previous OFFD as explained in the chapter 3 has an interaction between *MATLAB* and *FRED* from *Photon Engineering* (a commercial raytracer) [7] throughout the optimization which is explained using the figure 7.4. In order to establish this synchronous interaction without any overlap, delays t_1 and t_2 are manually introduced. All the operations excluding raytrace were implemented in the *MATLAB*. So the raytracer *FRED* waits t_2 seconds during such operations and *MATLAB* waits t_1 seconds during the raytrace vice-versa. This whole process continues until the desired functionality is attained. If the whole tool is present in the same program, the unnecessary writing of surfaces by the *MATLAB* in a defined location, *FRED* importing them and performing a raytrace, writing the evaluated results back for its access can be avoided. Writing on the hard disk and retrieving it each time is time intensive. To overcome this drawback, OFFD has to be completely implemented in one specific software, more obviously in a raytracer itself.

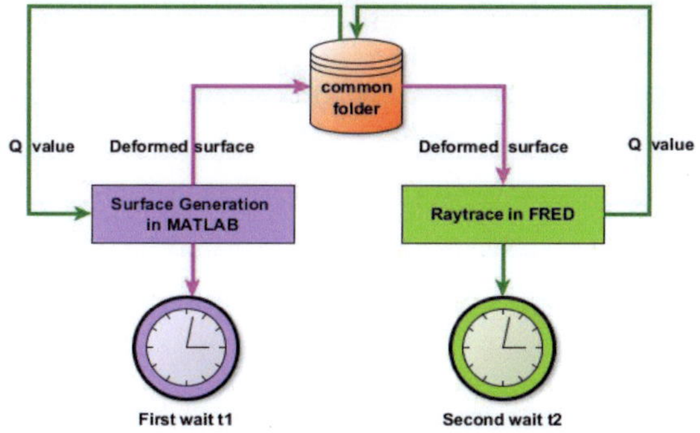

Figure 7.4: Setup showing the synchronization of *MATLAB* and raytracer *FRED* [7] for the sequential execution.

As OFFD relies heavily on the surface modification throughout its process, accurate software tools to manipulate the NURBS and to interact directly with it are required. Moreover, automation relies heavily on the script language provided by them. *Lucidshape* from

Synopsys [6] has all these features which make naturally a good choice to implement an entire OFFD in it. *Lucidshell*, the script language offered by them is the stripped version of C++ thereby enabling to write the OFFD using object oriented principles. This new OFFD is completely written in *Lucidshape* [6] thus eliminating the wait times and unnecessary file writings.

The proposed new architecture of the OFFD is as shown in the figure 7.5. Each block shown in green represent a specific functionality which has been implemented and analyzed as a part of this work. The pipeline consists of optimization block where Nelder Mead variants and genetic algorithm have been implemented and compared with each other. In the surface representation section, T-Splines have been added to perform local deformation in addition to the NURBS. The manufacturing tolerance analysis block does not allow surfaces that are not feasible to manufacture. The Preprocessing block helps to find the optimal grid points on its own. The merit functions help to evaluate the photometric performance of the surface. Finally, the raytracing block helps to characterize the optical surface where any tracing type can be selected. Study and improving the raytrace techniques are not a part of this dissertation and it relies completely on the commercial raytracer *Lucidshape* [6].

As each functional block stands on its own, it can be either turned on or off based on the requirements. For example, if the optical designer knows the best possible OFFD grid points well ahead, he can skip the preprocessing step by turning it off. If the application requires no surface analysis, then the surface analysis tool can be turned off. Moreover, new tools can be implemented and can be integrated to this pipeline. As more and more tools get added, the intelligence of this tool improves further which is elaborated in the outlook section.

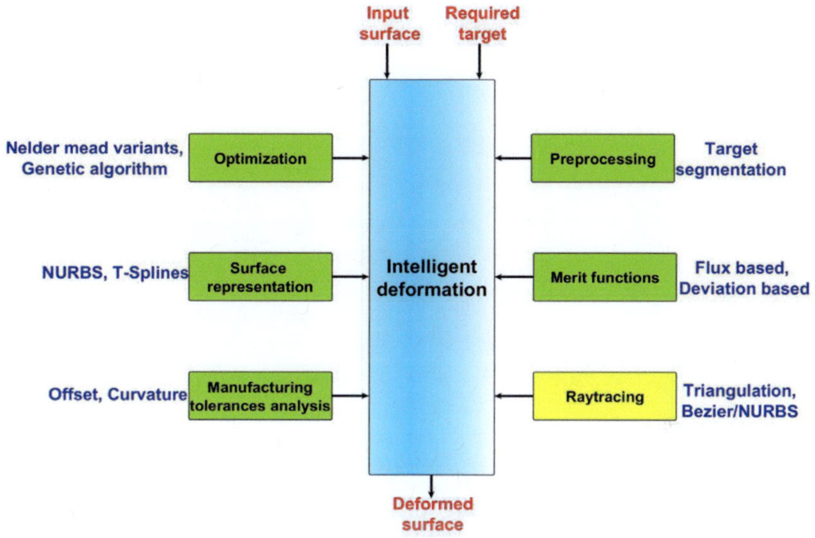

Figure 7.5: Software architecture of the intelligent OFFD.

7.4 COMPARISON OF PREVIOUS OFFD AND INTELLIGENT OFFD

In this section, comparison analysis is made between the previous and the intelligent OFFD in terms of its optical efficiency, speed, requirement of manual intervention and the comfort of the user by using street lighting lens and automotive stop lamps as design examples.

STREET LIGHTING LENS

The first step for the intelligent OFFD as seen in figure 7.3 is finding the best OFFD grid points using shift and scan principle. As this street lighting distribution is $y-$axis symmetric, it is sufficient to consider one-half of the grid. So the number of OFFD grid points for analysis is reduced to 18.

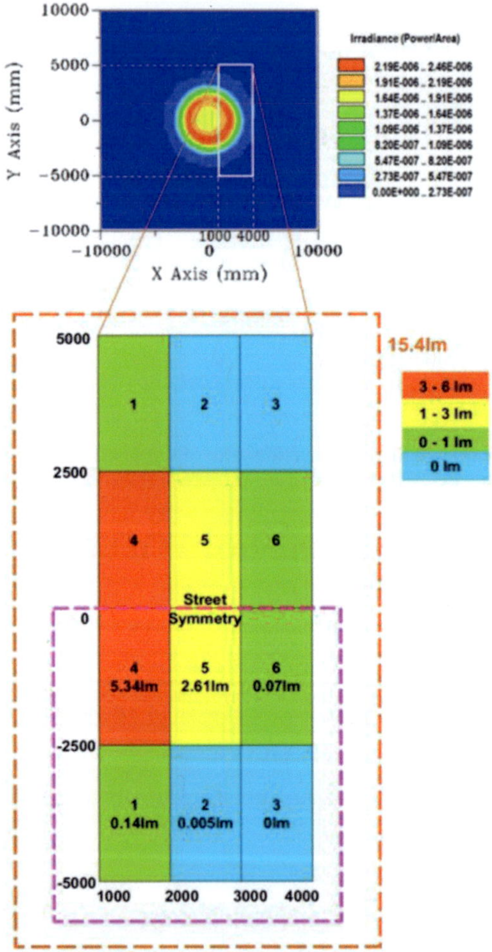

Figure 7.6: Illuminance distribution of the initial surface of the street lighting lens with the target area divided into 12 segments and the flux contribution at each segment.

Step 1 - Analysis on the distribution of the initial surface: To begin with, the light distribution is split into 12 small segments. Due to its symmetricity, only the lower half (highlighted using pink dotted lines) is considered and the segments have been marked as $1, 2, 3 \ldots, 6$ respectively as seen in the figure 7.6. The upper half is the replica of the lower one. From the figure 7.6., it is obvious that only 15.4% of the luminous flux emitted by the LED reaches the targeted region. In previous OFFD, this information is known but accurate analysis on estimating the amount of luminous flux at each segment was not carried out. From the figure 7.6., it is also clear that segments that are shown in turquoise (segment 2 and segment 3) require substantial improvement as there is almost no flux in the region. The segments in green (1 and 6) also require improvement.

Segment No.	OFFD grid points
1	4,5,6
2	1,4,5
3	3,12,15
4	1,14,15
5	1,2,5
6	1,2,5

Table 7.1: Results showing the relationship between OFFD grid points and each segment obtained after the shift and scan step

Step 2 - Shift and map the grid points: Once the segments are split and analyzed, the OFFD grid points that could improve these segments must be figured out. This can be found from the database where the mapped result between the grid and particular segment is stored. The results are shown in the table 7.1. The OFFD grid points that got mapped to the segments 1, 2, 3 and 6 found from the table 7.1 are $(1,2,3,4,5,6,12,15)$.

Step 3 - Optimization, target Analysis and re-iteration: The OFFD grid points $(1,2,3,4,5,6,12,15)$ found from the shift and scan step are given as optimization variables to the OFFD and optimization is carried out using deviation based merit function for 300 iterations.

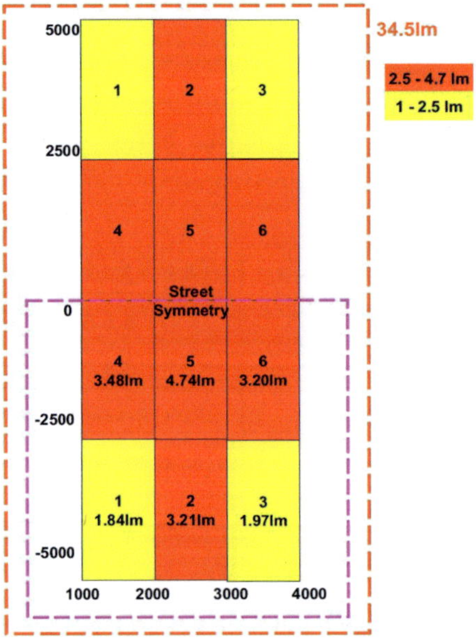

Figure 7.7: Illuminance distribution of the optimized surface after 300 iterations showing improved luminous flux.

After the optimization, the light distribution is examined for its improvement. From the figure 7.7, it is seen that the segments 1,2,3 and 6 have improved significantly. The luminous flux at the segment 5 has also found to be doubled though it is not intended. But this cannot be avoided because the OFFD grid points that are mapped to segment 6 have also been coupled with segment 5 as seen from the table 7.1. It is acceptable as far as the luminous flux in a particular segment do not exceed its tolerated limit. But there are certain cases, this is not acceptable and requires high uniformity. This can be attained by not selecting the mutual control points as optimization variables. This leads to higher uniformity in the light distribution but comes at an expense of optical efficiency.

This problem can be more intuitively solved using the following approach. The segments that are coupled with the same OFFD grid points are divided further and the relationship between the segmented region and grid points have to be established again. So now, the OFFD grid points are mutually exclusive and are mapped only to a particular segment. These grid points can be later selected as optimization variables. So different types of optical designer's requirements can be tweaked using this newly implemented method. The figure 7.7 also shows that the segments 1 and 3 have to be improved further. So the shift and scan step is carried out again with respect to the newly generated optical surface and a new map between the grid and the target is generated. The OFFD grid points that could improve segments 1 and 3 are found to be as $(2, 3, 6, 10, 11, 14)$. Then the surface is optimized again with these new OFFD grid points as optimization variables using flux based merit function for another 300 iterations. It is found that theses new set of optimization variables improved the illuminance distribution further as shown in the figure 7.8.

The optical efficiency of the distribution is around 45.8% leading to the same improvement of 30.4% as shown in the previous work. To attain this result, the intelligent OFFD required a total of 628 iterations which include the raytraces being carried during its preprocessing step to select the appropriate grid points. It took 4 minutes to perform preprocessing and 56 minutes for optimization which comes to a total of just an hour to complete this entire task in a 4 core 3.30GHz machine.

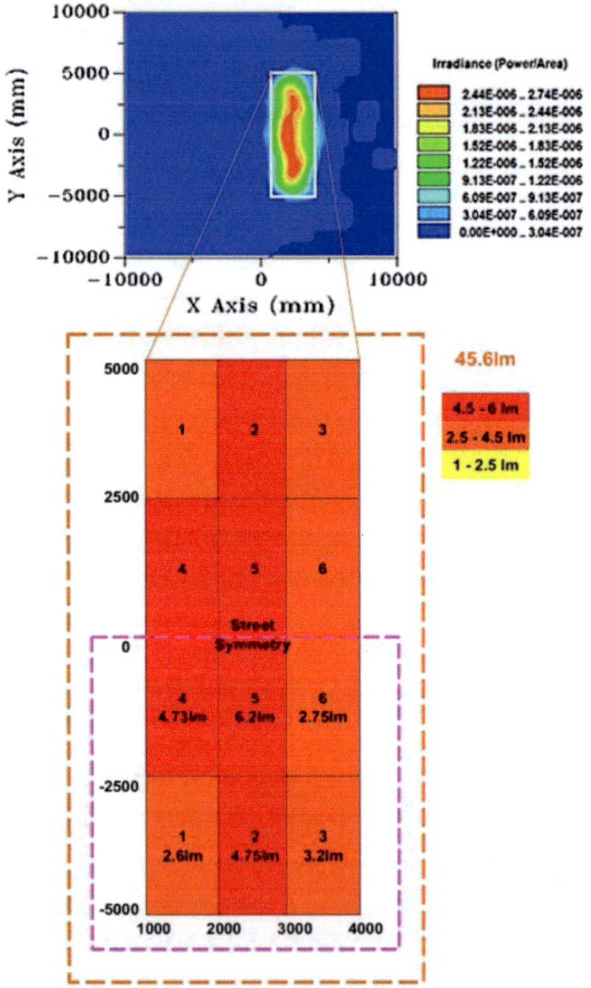

Figure 7.8: Illuminance distribution of the final optimized surface after 600 iterations showing improved luminous flux in each segment.

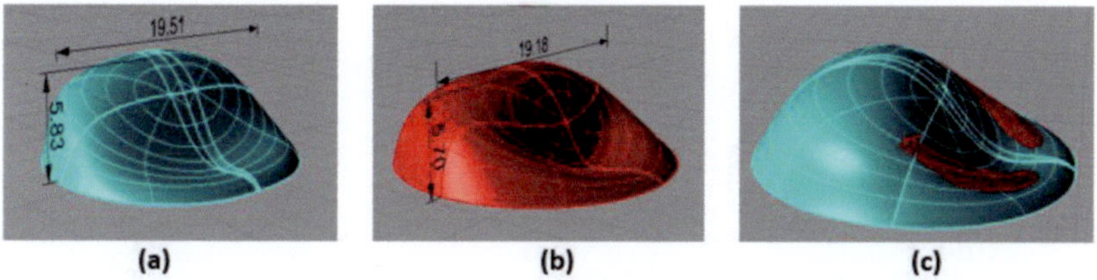

Figure 7.9: Deformed optical surface using (a) previous OFFD, (b) intelligent OFFD and (c) the false color representation showing a minimal change in shape between them with all dimension in millimeters.

The optical surface obtained as a result of the intelligent OFFD is shown in the figure 7.9. The optical surface generated using the intelligent OFFD is almost similar to the previous OFFD thereby proving that an intelligent OFFD could find the best possible optical surface at a faster rate autonomously using its imparted intelligence.

PERFORMANCE ANALYSIS

The optical efficiency and speed have been compared with the previous work [4] and is summarized in the table 7.2. From the table, it is evident that the intelligent OFFD could offer the same efficiency as that of the previous work. The previous work needed four different combinations of grid points with a total of 1500 iterations whereas the intelligent OFFD requires 628 iterations and just an hour to finish this task which is almost $209/60 = 3.5$ times faster than the previous OFFD for this particular task. The reason for its improvement in speed is due to the selection of the appropriate grid points using preprocessing technique and allowing only feasible surfaces to perform raytracing as explained in the chapter 6. More importantly, the new OFFD does not require any prior knowledge about the OFFD grid points and it finds them on the run which eliminates the manual intervention completely.

Parameters	Previous work [4]	Intelligent OFFD
Autonomous	No	Yes
Elimination of prior knowledge on grid	No	Yes
No.of iterations	1500	628
Grid combinations	$(1,3,4,6,14)$, (2), (5), $(2,3,11,12)$	$(1,2,3,4,5,6,12,15)$, $(2,3,6,10,11,14)$
Efficiency improvement ($\Delta\eta$)	30.2%	30.4%
Time required (minutes)	209	60

Table 7.2: Comparison chart showing the performance of the previous OFFD and the intelligent OFFD.

INTERESTING INSIGHTS

Though the objective of the intelligent OFFD is attained, there is always an inquisitiveness in understanding what happens to the optical surface if further optimization is carried out whether it improves further or remains same. So the whole process is repeated again and optimization is done in an attempt to improve the segments 1 and 6 as they have relatively less flux in comparison to others. Surprisingly, more luminous flux was redirected to this region but the flux from the middle of the light distribution (segment 5) was reduced drastically as seen in the figure 7.10. This is because the optimization algorithm tried hard to deform the surface to redirect more flux to segments 1 and 6. But the geometry of the optics is limited by the manufacturing tolerances in the OFFD. So when a new surface is created using the shifts provided by the optimization algorithm, the surface analysis tool rejected it and this continued for several iterations. Later, the optimization algorithm started to search differently and generated new shifts to the OFFD grid. The new shifts improved luminous flux in segments 1 and 6 but cannot deliver more flux to segment 5 anymore. This clearly showed that any light distribution is limited by the geometry of the optics which is popularly known as Etendue [1, 66].

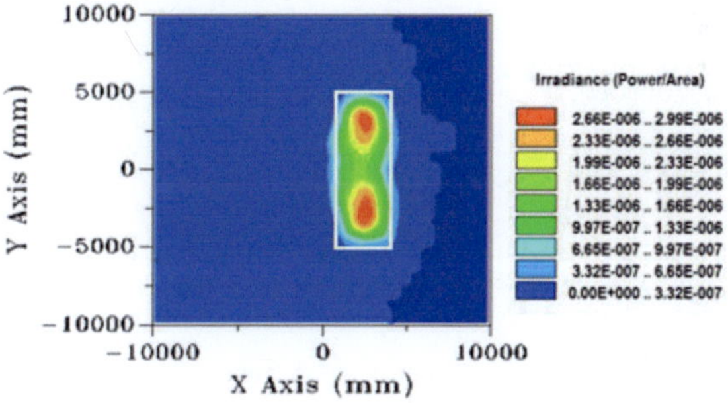

Figure 7.10: Illuminance distribution of the street light system when optimized beyond the physical limit.

AUTOMOTIVE STOP LAMP USING TIR HYBRID OPTICS

From the UNECE table 3.1, the intensity requirements for the stop lamp can be separated into two parts. The high-intensity requirements lie in the region -5° till +5° vertically and -10° till 10° horizontally and low intensity requirements lie at the outer edges -15° till +15° vertically and -45° till 45° horizontally. So the optimization has to be carried in two steps. The first one to redistribute the luminous flux optimally to the high-intensity region and the second one to obtain luminous flux in the low-intensity regions. The target high-intensity distribution needed for the automotive stop lamp is seen in the figure 7.11. When observed, it has both (horizontal and vertical) symmetries and so the OFFD grid points must be selected accordingly as it was already described in the section 7.2.2.

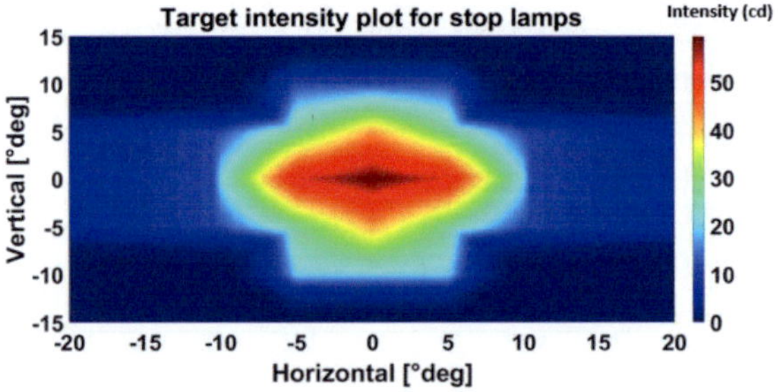

Figure 7.11: Target intensity distribution for stop lamp based on UNECE R7 S1 interpolated from the table 3.1.

Step 1 - Analysis on the distribution of the initial surface: As a first step, the intensity distribution produced by the initial TIR hybrid optics has to be analyzed. The distribution is collimated with high intensities around -2° till +2° with 810 cd in the main beam direction (0°). When calculated, 97% of the collected flux is present in the needed region which is obvious from the figure 7.12. So optimization based on flux based merit function will not help and hence deviation based merit function has to be used for such design tasks. To begin with, the intensity distribution is split into 8 segments marked as 1, 2, ..., 8 as shown in the figure 7.12. The segments are divided non-uniformly because of the uneven intensity requirements of the UNECE standard. The region with high intensity (in the main beam direction) are divided into small segments while the region that are far away from the middle with low intensity are divided as large segments. This uneven segmentation is done so that the luminous flux at each segment remains the same.

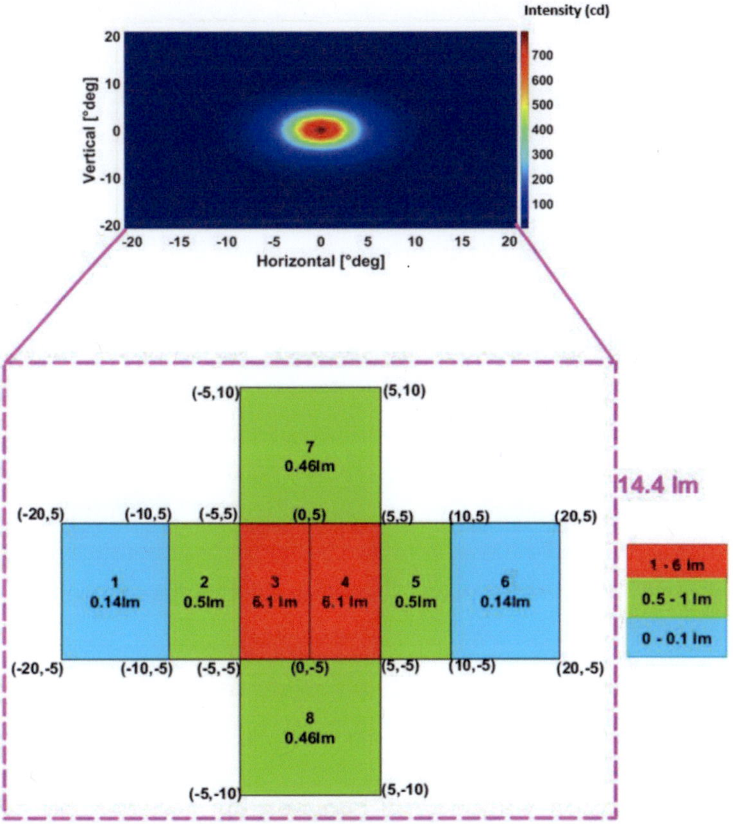

Figure 7.12: Intensity distribution of initial TIR hybrid optics with the target being split into 8 segments showing the flux contribution at each segment.

When the initial distribution is carefully analyzed, it is clear that segments that are shown in turquoise (segment 1 and segment 6) require substantial improvement as there is almost no flux in the region. The luminous flux in segments 3 and 4 represented in red has to be redirected away as they constitute more luminous flux than specified in the target requirements.

Step 2 - Shift and map the grid points: Once the segments are split and analyzed, the OFFD grid points that could improve these segments must be figured out using shift and scan approach. The grid points that got mapped to segments 1 and segment 6 using this approach are $(5, 11, 13, 14)$. These evaluated grid points will be given to the OFFD as optimization variables in the next step.

Step 3 - Optimization, target Analysis and re-iteration: Based on the selected OFFD grid points, optimization is carried out using deviation based merit function for 300 iterations. After the optimization, the light distribution is examined. From the figure 7.13, it is seen that the segments 1 and 6 have improved significantly. The luminous flux at the segments 3 and 4 have decreased by three times because the OFFD has redistributed the luminous flux from these segments to the rest of the segments.

As a second step, the deformed surface has to be optimized again to distribute the luminous flux to the outer edges -15° till +15° vertically and -45° till 45° horizontally. The same OFFD grid points are given as optimization variables and a short optimization routine consisting of 100 iterations using deviation mode merit function was sufficient as the intensity required in this region is only 0.3 cd.

After sufficient optimization and reiteration, the intensity at the middle $(0°)$ came around 60 cd which is exactly specified in the UNECE standard. The intensities at the other test points also fulfill the UNECE standard. The intensity at different test points for the deformed TIR hybrid optics is given in the table 7.3. The total luminous flux redirected to the high-intensity region (-5° till +5° vertically and -10° till 10° horizontally) was 11 lumens and the remaining luminous flux was utilized for the outer edges. So the total optical efficiency of the TIR hybrid optics as a whole is around 93.5%.

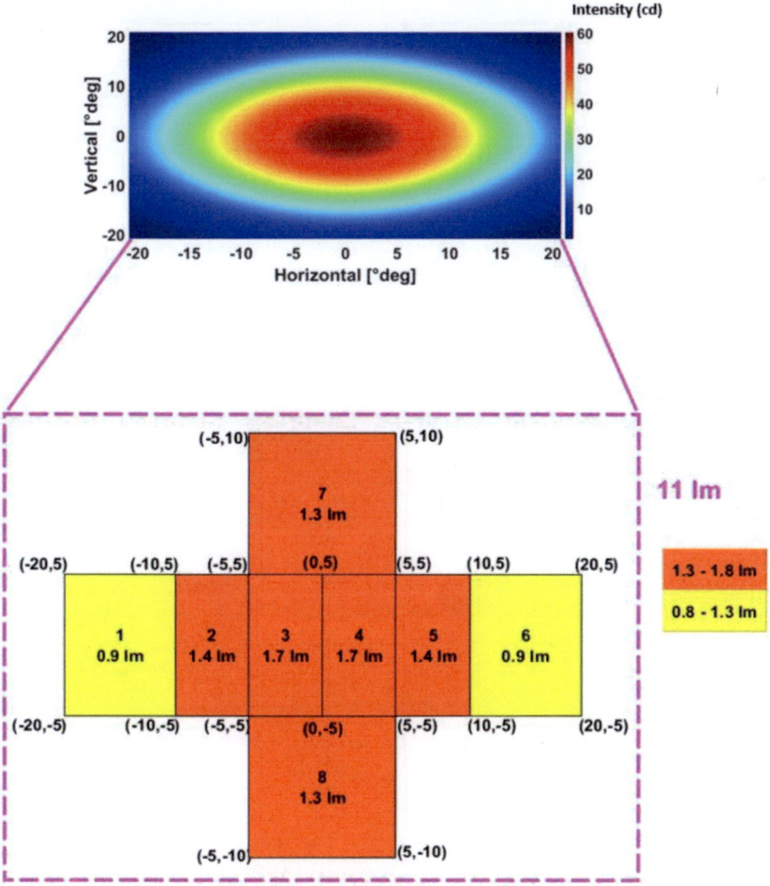

Figure 7.13: Intensity distribution of the optimized TIR hybrid optics showing improved luminous flux in each segment satisfying the UNECE requirements.

[*deg*]	-45°	-30°	-20°	-10°	-5°	0°	5°	10°	20°	30°	45°
15°	0.6	0.75	3	12	19	21	19	12	3	0.75	0.6
10°	0.45	-	-	-	37	40	37	-	-	-	0.45
5°	0.45	-	14	39	-	55	-	39	14	-	0.45
0°	0.5	-	-	45	57	64	57	45	-	-	0.5
-5°	0.45	-	14	39	-	55	-	39	14	-	0.45
-10°	0.45	-	-	-	37	40	37	-	-	-	0.45
-15°	0.6	0.75	3	12	19	21	19	12	3	0.75	0.6

Table 7.3: Luminous Intensity at test points obtained from the optimized TIR hybrid optics. The obtained light distribution has passed the UNECE R7 S1 at all the test points.

The TIR hybrid optics obtained as a result of the full automated intelligent OFFD (magenta) along with initial optics (as turquoise) are overlapped together as shown in the figure 7.14.

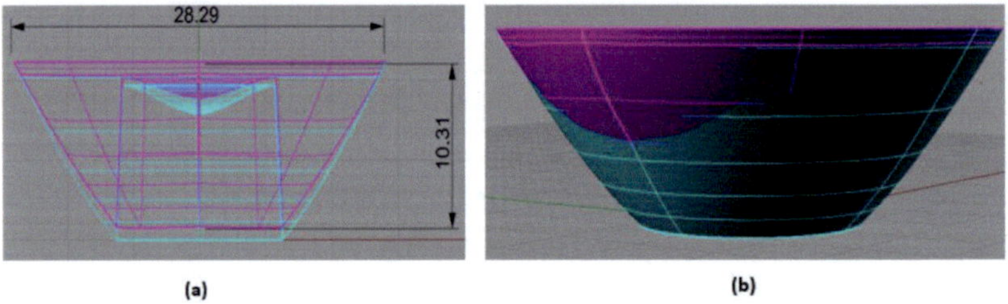

Figure 7.14: Initial (turquoise) and deformed (magenta) TIR hybrid optics overlapped to one another to show the difference in (a) front view and (b) perspective view with the dimensions in millimeters.

The deformation is more pronounced in the refracting surface of the TIR hybrid optics. In initial system, the light distribution is highly collimated and hence the refracting surface is convex. But in order to diverge the light to higher angles, the OFFD has modified the refracting surface. The thickness of the refracting surface has shrunken or flattened and has shifted forward. This means the refracting surface has taken the responsibility to redirect the flux to higher angles whereas the TIR surface redirects the flux to the middle. The side entry surface remains unaltered. The total contribution from the refracting and the TIR surface helps to fulfill the UNECE standard at the end.

7.5 CONCLUSION

This chapter showed how the OFFD could select the appropriate grid points by analyzing the illuminance or intensity distribution of the initial surface. Moreover, the OFFD has matured in such a way that all the algorithms and functionalities are hidden in a black box using object oriented principles. This implementation of the intelligent OFFD replaces the optical designer's efforts in terms of his decision making during the process. He needs to specify the target lighting requirements and the initial surface. The OFFD then searches the grid points and select the best possible ones. These grid points are specified as optimization variables to the OFFD. The surface is then optimized during which the quality of the surface is evaluated using manufacturing feasibility analysis tool and the photometric performance is analyzed using merit functions. At the end, a deformed surface satisfying the photometric needs and as well as feasible to manufacture will be created.

The comparison analysis between the previous work and the current OFFD was made for the street lighting lens. The current approach offered improvement in terms of speed and requires no manual intervention. Additionally, this method was also tested for automotive stop lamps. This involved multi surface deformation and it must be noted that the angular lighting requirements vary throughout the distribution. Still, the OFFD could perform well and the optimized TIR hybrid optics met the legal requirements.

Additionally, the initial surface selected for both the cases differ. For the street lighting system, the initial surface directs less luminous flux to the targeted area and the objective of the OFFD was to direct more flux to this region at high uniformity. But for the stop lamp, the initial surface already directed most of the flux within the angular region but not in a way the UNECE standard requires. But the intelligent OFFD could optimize and create an appropriate TIR hybrid optics satisfying the UNECE standard with high optical efficiency.

CHAPTER 8

SUMMARY

8.1 CONCLUSION

As a part of this dissertation, a new contribution has been made to the field of illumination optics by providing effective techniques to optimize freeform surfaces for extended sources. A new tool set has been developed with an aim to simplify the design and optimization process of the freeform optics. These methods had the potential to make the entire design process more efficient.The freeform deformation technique [28] was used as a backbone for optimization as it could provide global deformation with fewer optimization parameters.

Though NURBS surfaces are used commonly to represent freeform optics, there are certain drawbacks imposed by it which include difficulty in adding more control points in the needed region without change in its shape. So, an alternate way to represent the freeform surfaces known as T-Splines has been implemented in the OFFD to perform local deformation of optical surfaces. This was the first time that T-Splines have been implemented in the illumination optics system and its optical performance was analyzed. Results showed that the T-Splines could offer improvement in optical performance with less effort compared to NURBS when local modification of optical surfaces is required.

Once the surface has been accurately represented, the optimization starts using the cubic grid points as its optimization variables. To select optimal grid points, a suitable preprocessing technique was developed as a part of this dissertation. This involved establishing the relationship between the OFFD grid points and the target light distribution. Previously, as this relationship was not known, the OFFD relied heavily on the knowledge of an expert designer or trial and error approach. This work eliminated it by implementing a suitable solution. By doing so, the work of an optical designer or his intelligence had been transferred to this OFFD. This reduced the manual efforts manifold which means no knowledge regarding the OFFD grid surrounding the optical surface is required from the designer.

The selected optimal grid points are provided to the optimization algorithm in the next step. To improve this step, different variants of Nelder-Mead Simplex methods and genetic algorithm have been studied, implemented and compared against each other in terms of OFFD's perspective. Based on the exploratory study, different suggestions have been given to select a suitable algorithm based on the needs of the optical design task.

As the optimization begins, different surfaces are generated . But these surfaces are not always feasible to manufacture and raytracing such infeasible surfaces are computationally waste. So as a part of this dissertation, implementation of manufacturing feasibility analysis was carried out. The surfaces generated by the deformation algorithm were analyzed thoroughly for its curvature and its size. By using a penalty based non smooth merit function, raytracing was allowed only for the surfaces feasible to manufacture. The feasible surfaces are then photometrically evaluated and this process continues until the intelligent OFFD could find an optimal surface satisfying the photometric requirements and the geometric constraints.

As a result of this entire work, the overall speed was improved significantly due to the implemented design techniques. And most importantly, this entire operation does not require any human intervention in between the design process. The intelligent OFFD has grown in a way that when the initial optical surface and photometric requirements were specified as inputs, a surface satisfying the photometric requirements and feasible to manufacture will be generated at the end.

All the above-mentioned research techniques have been tested and compared with the previous work [4] by taking street lighting system as an example. Moreover, for concrete verification and to show the versatility and its potentiality of the proposed design techniques, another lighting application which is the automotive stop lamp design using TIR hybrid optics was taken. The TIR hybrid optics underwent several multi surface deformations as it had varying intensity requirements in its target distribution. Finally, it was optimized using the described techniques and was verified.

8.2 OUTLOOK

Though OFFD was improved to a greater extent, it has still more room to improve. This section gives an outlook on how this research could be further extended.

ETENDUE - A PHYSICAL LIMIT

The general trade-off in illumination optics always exists between the conservation of etendue (the product of emitting area times the solid angle of the emitted light cannot decrease without losses) and the mechanical or technological constraints. The most common effects caused by an extended source on the light distribution are smearing of the edges, forming peaks or dips at its center [10, 67]. The same problem of forming a dip at the middle and peaks at the edges was also addressed during the optimization of the street lighting system in chapter 7. The effects of extended sources in the light distribution can be studied using phase space mapping approach [68, 69]. But there is no reliable guideline so far that predicts the theoretically best performance of freeform optics for a given source and optics size for an arbitrary light distribution. When this is matured, it could give the physically attainable target distribution for particular source-optics requirements. Then this distribution can be fed into the OFFD to search a suitable surface.

Choice of Grid Shape

The employed deformation technique [28] in the OFFD is rapid and user-friendly to deform any object. But the cubic grid remained the same irrespective of the optical design tasks. When the shape of the grid is similar to the shape of the target, then the deformation process will be quicker as they are related to one another. But for complex grid shapes, establishing the relationship between the grid and the underlying NURBS is complex. Therefore only a few alternative simple grid shapes like cylinder and sphere are available [70]. This can be implemented in the OFFD and optical performance can be analyzed.

Machine Learning Techniques

To select suitable OFFD grid points for an arbitrary target distribution, the intelligent OFFD relies on target segmentation and shift and scan technique. Using these steps it could find the grid points automatically but these steps must be performed at the beginning of any optical design task. This can be improved using prediction, classification techniques found in machine learning. The machine learning helps to determine a mapping between input and output data, infer probability out of them and predict the output based on the past results [71, 72, 73]. Some of the techniques include neural networks, graphical models, decision tree learning approach, feedback based classification, support vector machines, Bayesian methods, principal component analysis and many others [74]. But in order to apply them in OFFD, these techniques have to be widely explored and a suitable technique must be picked and have to be implemented.

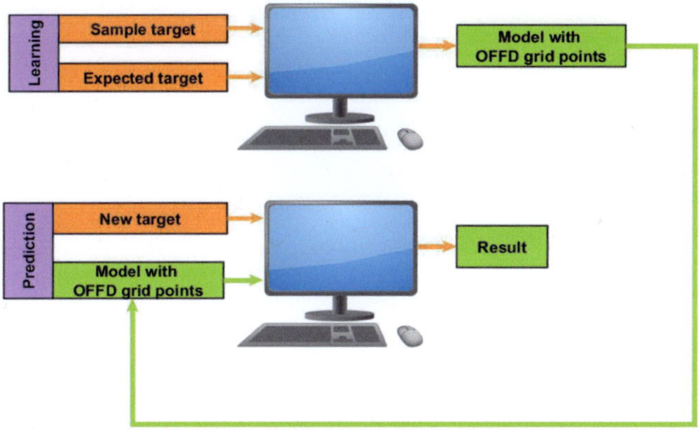

Figure 8.1: Simplified model of the OFFD with machine learning.

By doing so, the machine can be trained for various lighting distributions (in the order of thousands) to obtain a best possible grid points and this will be stored as a model. And when a new target distribution is given, it selects the best ones based on the prior information. This is faster but at an expense of high computational complexity. A simplified model of OFFD using machine learning approach is as shown in the figure 8.1.

Therefore, this dissertation improved the optimization using deformation technique in many aspects making it more advanced. As all these new techniques were incorporated in a modular fashion, it can be extended further. Due to its extensibility and its potentiality, this dissertation ends with a belief that it could open many doors in the future to add multiple features into the intelligent OFFD. This will make every illumination design task much easier and more interactive than today.

LIST OF REFERENCES

[1] R. J. Koshel, *Illumination Engineering: design with nonimaging optics.* John Wiley & Sons, 2012.

[2] A. Timinger, J. Unterhinninghofen, S. Junginger, and A. Hofmann, "Tolerancing free-form optics for illumination," in *SPIE Optical Systems Design*, International Society for Optics and Photonics, 2011.

[3] J. C. Miñano, P. Benítez, and A. Santamaría, "Free-form optics for illumination," *Optical review*, vol. 16, no. 2, pp. 99–102, 2009.

[4] S. Wendel, *Freiform-Optiken im Nahfeld von LEDs.* Ph.D Thesis, KIT Scientific Publishing, 2014.

[5] C. DeCusatis, *Handbook of applied photometry.* American Inst. of Physics, 1997.

[6] Synopsys LLC,"Lucidshape, an Automotive Lighting Design Software", https://optics.synopsys.com/lucidshape.

[7] Photon Engineering LLC, "FRED Optical Engineering Software", http://www.photonengr.com.

[8] H. Ries and J. Muschaweck, "Tailored freeform optical surfaces," *JOSA A*, vol. 19, no. 3, pp. 590–595, 2002.

[9] V. Oliker, "Mathematical aspects of design of beam shaping surfaces in geometrical optics," in *Trends in Nonlinear Analysis*, pp. 193–224, Springer, 2003.

[10] F. Fournier, *Freeform reflector design with extended sources.* Ph.D Thesis, University of Central Florida, 2010, http://purl.fcla.edu/fcla/etd/CFE0003311.

[11] P. Benı, J. C. Min, O. Dross, M. Herna, W. Falicoff, *et al.*, "Simultaneous multiple surface optical design method in three dimensions," *Optical Engineering*, vol. 43, no. 7, pp. 1489–1502, 2004.

[12] J. C. Miñano, P. Benítez, W. Lin, F. Muñoz, J. Infante, and A. Santamaría, "Overview of the sms design method applied to imaging optics," in *SPIE Optical Engineering+ Applications*, International Society for Optics and Photonics, 2009.

[13] J. C. Miñano, P. Benítez, J. Blen, and A. Santamaría, "High-efficiency free-form condenser overcoming rotational symmetry limitations," *Optics express*, vol. 16, no. 25, pp. 20193–20205, 2008.

[14] Z. Zhuang and F. Yu, "A contour calculation method for rapid freeform reflector construction with ellipsoid patches," *Optics & Laser Technology*, vol. 56, pp. 430–435, 2014.

[15] L. L. Doskolovich, E. A. Bezus, M. A. Moiseev, D. A. Bykov, and N. L. Kazanskiy, "Analytical source-target mapping method for the design of freeform mirrors generating prescribed 2d intensity distributions," *Optics express*, vol. 24, no. 10, pp. 10962–10971, 2016.

[16] W. Cassarly, "Iterative reflector design using a cumulative flux compensation approach," in *International Optical Design Conference*, Optical Society of America, 2010.

[17] A. S. Isaac and C. Neumann, "Design of compact freeform led flashlight capable of two different light distributions," in *SPIE Photonics Europe*, International Society for Optics and Photonics, apr 2016.

[18] F. R. Fournier, W. J. Cassarly, and J. P. Rolland, "Designing freeform reflectors for extended sources," in *SPIE Optical Engineering+ Applications*, International Society for Optics and Photonics, 2009.

[19] *Normungsausschuss Feuerwehrwesen (FNFW) im DIN*. DIN 14649 - Explosionsgeschuetzte Leuchten fuer Einsatzkraefte, 2011.

[20] W. J. Cassarly and M. J. Hayford, "Illumination optimization: the revolution has begun," in *International Optical Design Conference 2002*, pp. 258–269, International Society for Optics and Photonics, 2002.

[21] W. J. Cassarly, "Illumination merit functions," in *Optical Engineering+ Applications*, International Society for Optics and Photonics, 2007.

[22] S. Kudaev and P. Schreiber, "Automated optimization of non-imaging optics for luminaries," in *Optical Systems Design 2005*, International Society for Optics and Photonics, 2005.

[23] S. Kudaev and P. Schreiber, "Optical design & engineering optimizing led illumination systems," in *Proc. SPIE*, vol. 6342, 2007.

[24] R. J. Koshel, "Optimization of parameterized lightpipes," in *International Optical Design*, Optical Society of America, 2006.

[25] R. A. De Oliveira, M. F. de Medeiros Júnior, and R. F. A. Menezes, "Application of genetic algorithm for optimization on projects of public illumination," *Electric Power Systems Research*, vol. 117, pp. 84–93, 2014.

[26] G. F. Lima, J. Tavares, I. S. Peretta, K. Yamanaka, A. Cardoso, and E. Lamounier, "Optimization of lighting design usign genetic algorithms," in *Industry Applications (INDUSCON), 2010 9th IEEE/IAS International Conference on*, pp. 1–6, IEEE, 2010.

[27] R. J. Koshel, "Fractional optimization of illumination optics," in *Optical Engineering+ Applications*, International Society for Optics and Photonics, 2008.

[28] T. W. Sederberg and S. R. Parry, "Free-form deformation of solid geometric models," *ACM SIGGRAPH computer graphics*, vol. 20, no. 4, pp. 151–160, 1986.

[29] R. Raffin, "Free form deformations or deformations non-constrained by geometries or topologies," in *Deformation Models*, pp. 49–74, Springer, 2013.

[30] J. Nocedal and S. Wright, *Numerical Optimization*. Springer Verlag, 1999.

[31] *Product Family Datasheet: Cree XLamp XP-G2 LEDs*. CREE, 2012.

[32] *Uniform provisions concerning the approval of front and rear position lamps, stop-lamps and end-outline marker lamps for motor vehicles and their trailers*. UN Vehicle Regulations - 1958 Agreement, Addendum 6: Regulation No. 7, November 2012.

[33] *Technical Datasheet: LUXEON Rebel LEDs*. Philips LUMILEDS, 2015.

[34] R. J. Koshel, "Simplex optimization method for illumination design," *Optics letters*, vol. 30, no. 6, pp. 649–651, 2005.

[35] J. A. Nelder and R. Mead, "A simplex method for function minimization," *The computer journal*, vol. 7, no. 4, pp. 308–313, 1965.

[36] R. J. Koshel, "Enhancement of the downhill simplex method of optimization," in *International Optical Design Conference*, Optical Society of America, 2002.

[37] R. R. Barton and J. S. Ivey Jr, "Nelder-mead simplex modifications for simulation optimization," *Management Science*, vol. 42, no. 7, pp. 954–973, 1996.

[38] F. Gao and L. Han, "Implementing the nelder-mead simplex algorithm with adaptive parameters," *Computational Optimization and Applications*, vol. 51, no. 1, pp. 259–277, 2012.

[39] S.-K. S. Fan and E. Zahara, "A hybrid simplex search and particle swarm optimization for unconstrained optimization," *European Journal of Operational Research*, vol. 181, no. 2, pp. 527–548, 2007.

[40] P. C. Wang and T. E. Shoup, "Parameter sensitivity study of the nelder–mead simplex method," *Advances in Engineering Software*, vol. 42, no. 7, pp. 529–533, 2011.

[41] R. L. Haupt and S. E. Haupt, *Practical genetic algorithms*. John Wiley & Sons, 2004.

[42] K. E. Moore, "Algorithm for global optimization of optical systems based upon genetic competition," in *Proc. SPIE*, vol. 3780, pp. 40–47, 1999.

[43] C. Darwin, "On the origin of species by means of natural selection," *John Murray, London*, 1859.

[44] A. Alajmi and J. Wright, "Selecting the most efficient genetic algorithm sets in solving unconstrained building optimization problem," *International Journal of Sustainable Built Environment*, vol. 3, no. 1, pp. 18–26, 2014.

[45] O. Boyabatli and I. Sabuncuoglu, "Parameter selection in genetic algorithms," *Journal of Systemics, Cybernetics and Informatics*, vol. 4, no. 2, p. 78, 2004.

[46] B. L. Miller, D. E. Goldberg, *et al.*, "Genetic algorithms, tournament selection, and the effects of noise," *Complex systems*, vol. 9, no. 3, pp. 193–212, 1995.

[47] L. Piegl, "Why nurbs?," *IEEE Computer Graphics and Applications*, pp. 55–71, 1991.

[48] T. W. Sederberg, D. L. Cardon, G. T. Finnigan, N. S. North, J. Zheng, and T. Lyche, "T-spline simplification and local refinement," in *ACM transactions on graphics (TOG)*, vol. 23, pp. 276–283, ACM, 2004.

[49] D. L. Cardon, "T-spline simplification," *Brigham Young University-Provo*, 2007,http://scholarsarchive.byu.edu/etd/871.

[50] E. Bailey and S. Carayon, "Beyond nurbs: enhancement of local refinement through t-splines," in *Optical Engineering+ Applications*, International Society for Optics and Photonics, 2007.

[51] W. Boehm, "Inserting new knots into b-spline curves," *Computer-Aided Design*, vol. 12, no. 4, pp. 199–201, 1980.

[52] E. Cohen, T. Lyche, and R. Riesenfeld, "Discrete b-splines and subdivision techniques in computer-aided geometric design and computer graphics," *Computer graphics and image processing*, vol. 14, no. 2, pp. 87–111, 1980.

[53] T. Lyche and K. Mørken, "Knot removal for parametric b-spline curves and surfaces," *Computer Aided Geometric Design*, vol. 4, no. 3, pp. 217–230, 1987.

[54] T. W. Sederberg, J. Zheng, A. Bakenov, and A. Nasri, "T-splines and t-nurccs," in *ACM transactions on graphics (TOG)*, vol. 22, pp. 477–484, ACM, 2003.

[55] L. R. Seidenberg, R. B. Jerard, and J. Magewick, "Surface curvature analysis using color," in *Visualization, 1992. Visualization'92, Proceedings., IEEE Conference on*, pp. 260–267, IEEE, 1992.

[56] J.-H. Chuang and C. M. Hoffmann, "Curvature computations on surfaces in n-space," *ESAIM: Mathematical Modelling and Numerical Analysis*, vol. 26, no. 1, pp. 95–112, 1992.

[57] G. Farin and N. Sapidis, "Curvature and the fairness of curves and surfaces," *IEEE Computer Graphics and Applications*, vol. 9, no. 2, pp. 52–57, 1989.

[58] A. Knoebel, R. Laubenbacher, J. Lodder, and D. Pengelley, *Mathematical masterpieces: Further chronicles by the explorers*. Springer Science & Business Media, 2007.

[59] M. Garman and J. Bonnie, "Curvature of surfaces in 3-space," *Verge*, 2010.

[60] H. Hagen, S. Hahmann, and T. Schreiber, "Visualization and computation of curvature behaviour of freeform curves and surfaces," *Computer-Aided Design*, vol. 27, no. 7, pp. 545–552, 1995.

[61] A. N. Pressley, *Elementary differential geometry*. Springer Science & Business Media, 2010.

[62] A. Roberts, "Curvature attributes and their application to 3 d interpreted horizons," *First break*, vol. 19, no. 2, pp. 85–100, 2001.

[63] S. Musuvathy, E. Cohen, J. Damon, and J.-K. Seong, "Principal curvature ridges and geometrically salient regions of parametric b-spline surfaces," *Computer-Aided Design*, vol. 43, no. 7, pp. 756–770, 2011.

[64] W. Boehm, "Differential geometry ii," *Curves and Surfaces for Computer Aided Geometric Design: A Practical Guide*, pp. 267–83, 1990.

[65] L. Mohammed and G. Al-Kindi, "An approach to 3d-surface curvature analysis," *Engg and Technology Journal*, vol. 24, no. 7, pp. 844–852, 2005.

[66] A. Isaac, J. Long, and C. Neumann, "Optimization of freeform optics using t-splines in led illumination design," in *LED professional Symposium+Expo*, 2017.

[67] A. Isaac, S. Wendel, and C. Neumann, "Analysis on the effects of optics-source scaling for the free-form deformation optimization," in *Licht , The Hague*, 2014.

[68] D. Rausch, M. Rommel, A. M. Herkommer, and T. Talpur, "Illumination design for extended sources based on phase space mapping," *Optical Engineering*, vol. 56, no. 6, 2017.

[69] A. M. Herkommer and D. Rausch, "Phase space transformations-a different way to understand freeform optical systems," in *Freeform Optics*, Optical Society of America, 2013.

[70] S. Coquillart, *Extended free-form deformation: a sculpturing tool for 3D geometric modeling*, vol. 24. ACM, 1990.

[71] J. Thrane, J. Wass, M. Piels, J. C. Diniz, R. Jones, and D. Zibar, "Machine learning techniques for optical performance monitoring from directly detected pdm-qam signals," *Journal of Lightwave Technology*, vol. 35, no. 4, pp. 868–875, 2017.

[72] D. Zibar, M. Piels, R. Jones, and C. G. Schäeffer, "Machine learning techniques in optical communication," *Journal of Lightwave Technology*, vol. 34, no. 6, pp. 1442–1452, 2016.

[73] L. Waller and L. Tian, "Computational imaging: Machine learning for 3d microscopy," *Nature*, vol. 523, no. 7561, pp. 416–417, 2015.

[74] K. P. Murphy, *Machine learning: a probabilistic perspective.* MIT press, 2012.

Appendix A

Nelder-Mead Simplex method

For better understanding of the Nelder-Mead simplex method, an example of how a 2D simplex works is explained in this section. The figureA.1 shows how the simplex is performed for the three vertices x_1, x_2 and x_3.

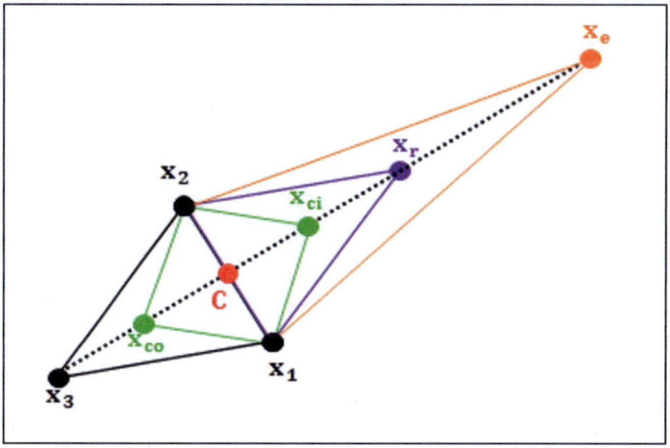

Figure A.1: An example of 2-dimensional simplex

A.1 Initial Calculation

The algorithm begins by evaluating the initial system having the variables $x_1, x_2, \ldots, x_n, x_{n+1}$.

SORTING

The result is then ordered: $f(x_1) < f(x_2) < \cdots < f(x_n) < f(x_{n+1})$. We define x_1 as the best vertex and x_{n+1} as the worst vertex such that

$$f(x_1) = min\left(\sum_{i=1}^{n+1} f(x_i)\right) \tag{A.1}$$

$$f(x_{n+1}) = max\left(\sum_{i=1}^{n+1} f(x_i)\right) \tag{A.2}$$

COMPUTATION OF CENTROID

The centroid C is calculated for all the points except the vertex with the highest evaluated value x_{n+1} as given in the equationA.3

$$C = \frac{\sum_{i=1}^{n} x_i}{n} \tag{A.3}$$

A.2 REFLECTION

The reflection point denoted as x_r is found by reflecting the highest value x_{n+1} through the centroid C using the relation in equationA.4

$$x_r = C + \alpha (C - x_{n+1}) \tag{A.4}$$

where α is the reflection parameter. Then the function is evaluated at this reflected point: $f_r = f(x_r)$. If f_r lies between the function value of best vertex $f(x_1)$ and the function value of second worst vertex $f(x_n)$, i.e., $f(x_1) < f(x_r) < f(x_n)$, replace the worst vertex x_{n+1} by the reflection point x_r.

A.3 EXPANSION

If the function evaluated at the reflection point $f(x_r)$ is smaller than the value of best vertex , i.e., $f(x_r) < f(x_1)$, then compute the expansion point x_e using the relation in equationA.5

$$x_e = C + \gamma (x_r - C) \tag{A.5}$$

where γ is the expansion parameter. If $f(x_e) < f(x_r)$, replace the worst vertex x_{n+1} by x_e , otherwise replace x_{n+1} by x_r .

A.4 CONTRACTION

If $f(x_r) > f(x_{n+1})$, compute inside contraction point x_{ci} with the relation given in equationA.6

$$x_{ci} = C + \beta(x_{n+1} - C) \tag{A.6}$$

If $f(x_r) < f(x_{n+1})$, compute outside contraction point x_{co} using the relation in equationA.7

$$x_{co} = C + \beta(x_{n+1} + C) \tag{A.7}$$

where β in equations is the contraction parameter. If $f(x_{ci}) < f(x_{n+1})$ or $f(x_{co}) < f(x_{n+1})$, replace x_{n+1} by x_{ci} or x_{co} respectively.

A.5 SHRINK

If$f(x_{ci}) > f(x_{n+1})$ or $f(x_{co}) > f(x_{n+1})$,replace all the vertices x_i , $i \in [2, n+1]$ using the relation shown in equationA.8

$$x_i = x_1 + \rho(x_i - x_1) \tag{A.8}$$

where ρ is the shrink parameter.

APPENDIX B

KNOT INSERTION AND BASIS FUNCTION REFINEMENT

The main advantage of the T-splines is that the surface holds its shape even when a new control point is inserted. This section shows how this could be achieved using an example. As seen in the chapter, the refinement of the basis function $N_i\,(u,v)$ can be simplified into two individual uni-variate basis functions $N_{ui}\,(u,v)$ and $N_{vi}\,(u,v)$ respectively. This section shows the refinement of one of the uni-variate function $N_{ui}\,(u,v)$.

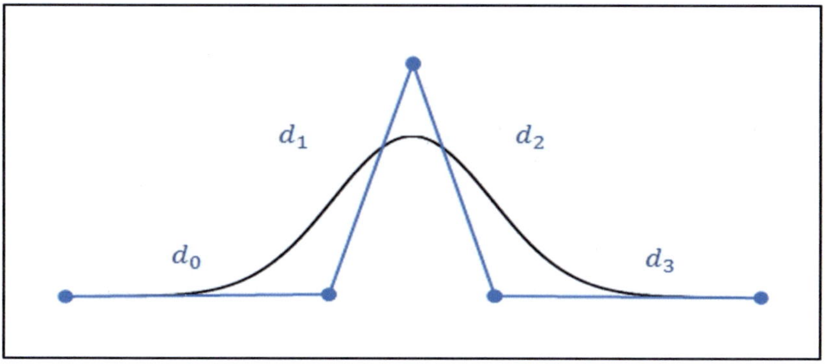

Figure B.1: B-spline basis function with knot intervals (d_0, d_1, d_2, d_3) and control point ordinates $(0,0,1,0,0)$ from left to right before refinement

B.1 REFINEMENT OF THE UNI-VARIATE FUNCTION

The example selected is just a simple curve containing a peak at the middle and zero for the rest of the ordinates. So this could be approximated as a B-spline curve as shown in the figureB.1. so the control point ordinate values are given from left to right as $(0, 0, 1, 0, 0)$ with the knot intervals (d_0, d_1, d_2, d_3). An insertion to basis function leads to two scaled basis functions. So the sum of these two scaled functions equals to the original basis function without any insertion and can be written as [49]:

$$N_{ui}(u) = c_1 N_{ui1}(u) + c_2 N_{ui2}(u) \tag{B.1}$$

where c_1 and c_2 are the so called control point ordinate values. In our example, we are inserting a new point in the knot interval d_1. First, the d_1 must be split into two segments d_{1L} and d_{1R} as shown in the figureB.2.Then these new intervals and interval neighbors are mapped onto the control polygon as shown in the figureB.3. For instance, the interval d_2 needs to be mapped like the previous neighbor d_1 by splitting into d_{1L} and d_{1R} and its next neighbor d_3.

The figureB.4 shows the uni-variate basis function after refinement. Two scaled basis functions $N_{ui1}(u)$ and $N_{vi1}(u)$ are associated with knot intervals $(d_0, d_{1L}, d_{1R}, d_2)$ and $(d_{1L}, d_{1R}, d_2, d_3)$. c_1 and c_2 are the new positions of new ordinates. Therefore the refined basis function has new control points represented from left to right as $(0, 0, c_1, c_2, 0, 0)$ where $c_1$and c_2 are found using

$$c_1 = \frac{d_0 + d_{1L}}{d_0 + d1 + d_2} \tag{B.2}$$

$$c_2 = \frac{d_{1R} + d_2 + d_3}{d_0 + d1 + d_2} \tag{B.3}$$

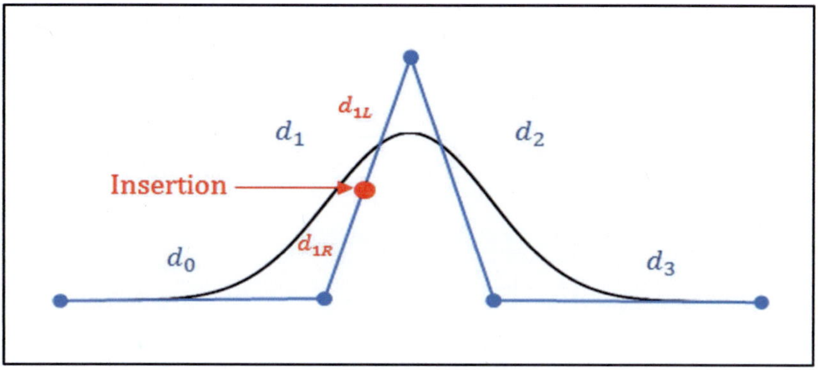

Figure B.2: B-spline basis function showing a split in the interval d_1 in to two segments d_{1L} and d_{1R} when a new knot is inserted

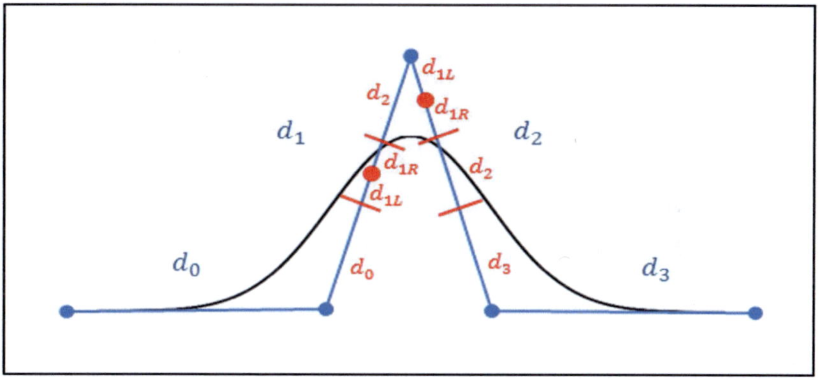

Figure B.3: B-spline basis function showing how the knot intervals are getting modified and mapped

In this example, the expressions c_1 and c_2 are derived when the interval d_1 was split. In the same manner, the expressions can be derived for splits of the other remaining intervals too and this is summarized in the tableB.1.

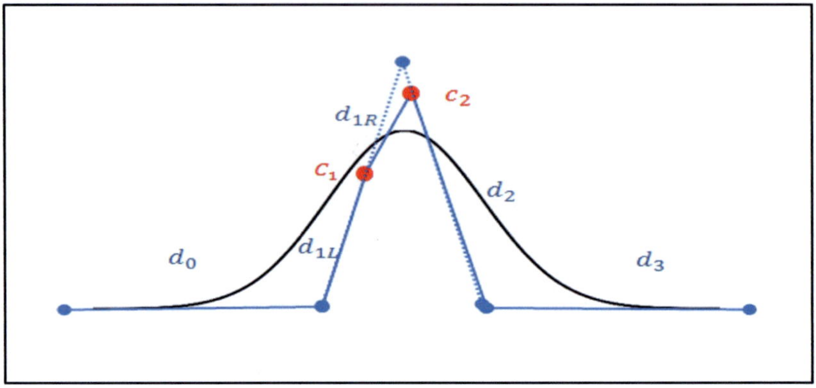

Figure B.4: The refined B-spline basis function showed in solid line

split interval	c_1	c_2	knot intervals after refinement
d_0	$\frac{d_{0L}}{d_0+d_1+d_2}$	1	$N_{ui1}(u) : (d_{0L}, d_{0R}, d_1, d_2)$ $N_{ui2}(u) : (d_{0R}, d_1, d_2, d_3)$
d_1	$\frac{d_0+d_{1L}}{d_0+d_1+d_2}$	$\frac{d_{1R}+d_2+d_3}{d_1+d_2+d_3}$	$N_{ui1}(u) : (d_0, d_{1L}, d_{1R}, d_2)$ $N_{ui2}(u) : (d_{1L}, d_{1R}, d_2, d_3)$
d_2	$\frac{d_0+d_1+d_{2L}}{d_0+d_1+d_2}$	$\frac{d_{2R}+d_3}{d_1+d_2+d_3}$	$N_{ui1}(u) : (d_0, d_1, d_{2L}, d_{2R})$ $N_{ui2}(u) : (d_{2L}, d_{2R}, d_2, d_3)$
d_3	1	$\frac{d_R}{d_1+d_2+d_3}$	$N_{ui1}(u) : (d_0, d_1, d_2, d_{3L})$ $N_{ui2}(u) : (d_1, d_2, d_{3L}, d_{3R})$

Table B.1: Ordinate values and knot intervals after refinement when original intervals d_0, d_1, d_2, d_3 are split

List of Acronyms

Abbreviation	Explanation
2D	Two-Dimensional
3D	Three-Dimensional
CAD	Computer Aided Design
CAGD	Computer Aided Graphics Design
CAL	Computer Aided Lighting
COB	Chip on Board
CIE (French)	Commission Internationale de l'eclairage
FFD	Freeform Deformation
FRED	name of a commercial raytracing program
FOM	Figure-of-Merit
FWHM	Full Width Half Maximum
LED	Light emitting diode
NURBS	Nonuniform Rational B-Spline
NP	Nondeterministic Polynomial
OFFD	Optimization using Freeform Deformation
PDE	Partial Differential Equation
PMMA	Polymethyl methacrylate
RMS	Root Mean Square
SMS	Simultaneous Multiple Surfaces
TIR	Total Internal Reflection
UNECE	United Nations Economics Commission of Europe

List of Symbols

Symbol	Definition
η	efficiency of an optical system
ϕ_{source}	luminous flux from the source $[lm]$
ϕ_{target}	luminous flux at the target $[lm]$
α	reflection parameter in simplex
γ	expansion parameter in simplex
β	contraction parameter in simplex
ρ	shrink parameter in simplex
Q_{flux}	figure of merit on the luminous flux at the target
Q_{dev}	figure of merit for the deviation between distributions
ΔQ_{dev}	improvement by the deviation merit function
$\Delta\eta$	improvement in efficiency
κ	Gaussian curvature
K_{max}	maximum principal curvature
K_{min}	minimum principal curvature
$n_{critical}$	total number of critical points for surface analysis
σ	variance in the Gaussian distribution
Q_{pho}	photometric figure of merit
Q_{geo}	figure of merit on geometric measure
μ_{geo}	penalty parameter for the geometric measure
μ_{pho}	penalty parameter for the photometric measure
δ	step size used in the optimization algorithm
$Curv_{max}$	maximum principal surface curvature
$Curv_{min}$	minimum principal surface curvature

Publications by the Author

- A. S. Isaac, S.Wendel and C. Neumann, "Analysis on the effects of optics-source scaling for the free-form deformation optimization," in Licht, The Hague, sep 2014.
- A. S. Isaac and C. Neumann, "Design of compact freeform LED flashlight capable of two different light distributions," in SPIE Photonics Europe, pp. 98890X–98890X, International Society for Optics and Photonics, apr 2016 (Best paper award).
- A. S. Isaac and C. Neumann, "Optimization of freeform optics using T-splines in LED illumination design," in LED professional Symposium+Expo, sep 2017 (nominated for the scientific award).
- A.S. Isaac and C. Neumann, "Optimization of freeform surfaces using intelligent deformation techniques for LED applications," in Advanced Optical Technologies, Vol. 7, Issue 1, 2018 (submitted and under review).

ACKNOWLEDGMENTS

I would like to thank my advisor Prof. Dr. Cornelius Neumann who gave me an opportunity to carry out the research in the field of illumination optics at the Light Technology Institute which was completely new to me. Wholeheartedly, I would like to thank for the trust he laid on me in not only on the doctoral work but also the projects which we worked together. The discussions, thoughtful ideas, honest feedback have shaped me to the person where I am today.

Next I would like to thank Prof. Dr. Wilhelm Stork for his kind acceptance as an examiner for my doctoral examination.

I acknowledge Dr. Simon Wendel who helped me to dive into the ocean of freeform optics. The help extended to me in my initial days to get familiarize in this field and his knowledge what he has transferred to me can never be forgettable.

I express my deep gratitude to Dr. Christian Herbold, Dr. Simon Wendel and Patric Jahn for carefully reading this manuscript and providing many constructive feedback. I thank all my colleagues in our working group for the time we spent together in industrial collaborated projects and times of fun.

This work would not be possible without the support from Karlsruhe School of Optics and Photonics by providing scholarship for the initial two years until me being accepted as a scientific staff by the KIT. I would also thank the Hochschulgemeinschaft für Lichttechnik for the financial aid in publishing my dissertation.

SPEKTRUM DER LICHTTECHNIK

Lichttechnisches Institut Karlsruher Institut für Technologie (KIT)
ISSN 2195-1152

Band 1	Christian Jebas **Physiologische Bewertung aktiver und passiver Lichtsysteme im Automobil.** 2012 ISBN 978-3-86644-937-4
Band 2	Jan Bauer **Effiziente und optimierte Darstellungen von Informationen auf Grafikanzeigen im Fahrzeug.** 2013 ISBN 978-3-86644-961-9
Band 3	Christoph Kaiser **Mikrowellenangeregte quecksilberfreie Hochdruckgasentladungslampen.** 2013 ISBN 978-3-7315-0039-1
Band 4	Manfred Scholdt **Temperaturbasierte Methoden zur Bestimmung der Lebensdauer und Stabilisierung von LEDs im System.** 2013 ISBN 978-3-7315-0044-5
Band 5	André Domhardt **Analytisches Design von Freiformoptiken für Punktlichtquellen.** 2013 ISBN 978-3-7315-0054-4
Band 6	Franziska Herrmann **Farbmessung an LED-Systemen.** 2014 ISBN 978-3-7315-0173-2
Band 7	Simon Wendel **Freiform-Optiken im Nahfeld von LEDs.** 2014 ISBN 978-3-7315-0251-7
Band 8	Carmen Kettwich **Ablenkung im Straßenverkehr und deren Einfluss auf das Fahrverhalten.** 2014 ISBN 978-3-7315-0288-3

Die Bände sind unter www.ksp.kit.edu als PDF frei verfügbar oder als Druckausgabe bestellbar.

Band 9 Steffen Michenfelder
Konzeption, Realisierung und Verifikation eines automobilen Forschungsscheinwerfers auf Basis von Digitalprojektoren. 2015
ISBN 978-3-7315-0301-9

Band 10 Celal Mohan Ögün
Surface wave driven molecular low pressure plasmas for general lighting. 2016
ISBN 978-3-7315-0464-1

Band 11 Theresa Bonenberger
LED Farbmischung mit chaotischen Lichtleitern. 2016
ISBN 978-3-7315-0480-1

Band 12 Michael Schöne
Diffraktive Optiken im Automobil: Achromatisierung, Athermalisierung, Formung von Scheinwerferlichtverteilungen. 2017
ISBN 978-3-7315-0613-3

Band 13 Tobias Werner
Simulation, Aufbau und Charakterisierung von autostereoskopischen Display-Systemen im Fahrzeugbereich. 2017
ISBN 978-3-7315-0617-1

Band 14 Christian Herbold
Entwicklung und Herstellung naturähnlich verzweigter Kühlkörper für LED-Systeme. 2017
ISBN 978-3-7315-0635-5

Band 15 Carsten Gut
Laserbasierte hochauflösende Pixellichtsysteme. 2018
ISBN 978-3-7315-0710-9

Band 16 Annie Shalom Samji Isaac Chandra
Intelligent Freeform Deformation for LED Illumination Optics. 2018
ISBN 978-3-7315-0741-3